Boolean Reasoning

Boolean Reasoning

The Logic of Boolean Equations

by

Frank Markham Brown
Air Force Institute of Technology

Kluwer Academic Publishers
Boston/Dordrecht/London

Distributors for North America:
Kluwer Academic Publishers
101 Philip Drive
Assinippi Park
Norwell, Massachusetts 02061, USA

Distributors for all other countries:
Kluwer Academic Publishers Group
Distribution Centre
Post Office Box 322
3300 AH Dordrecht, THE NETHERLANDS

Consulting Editor: Jonathan Allen

Library of Congress Cataloging-in-Publication Data

Brown, Frank Markham, 1930–
    Boolean reasoning / by Frank Markham Brown.
        p.   cm.
    Includes bibliographical references (p. 247-264).
    Includes index.
    ISBN 0-7923-9121-7
    1. Algebra, Boolean.   I. Title.
QA10.3.B76   1990
511.3 '24—dc20   90–4714
                                                            CIP

Contents

Preface

This book is about the logic of Boolean equations. Such equations were central in the "algebra of logic" created in 1847 by Boole [12, 13] and developed by others, notably Schröder [178], in the remainder of the nineteenth century. Boolean equations are also the language by which digital circuits are described today.

Logicians in the twentieth century have abandoned Boole's equation-based logic in favor of the more powerful predicate calculus. As a result, digital engineers—and others who use Boole's language routinely—remain largely unaware of its utility as a medium for reasoning. The aim of this book, accordingly, is to is to present a systematic outline of the logic of Boolean equations, in the hope that Boole's methods may prove useful in solving present-day problems.

Two Logical Languages

Logic seeks to reduce reasoning to calculation. Two main languages have been developed to achieve that object: Boole's "algebra of logic" and the predicate calculus. Boole's approach was to represent classes (*e.g.,* happy creatures, things productive of pleasure) by symbols and to represent logical statements as equations to be solved. His formulation proved inadequate, however, to represent ordinary discourse. A number of nineteenth-century logicians, including Jevons [94], Poretsky [159], Schröder [178], Venn [210], and Whitehead [212, 213], sought an improved formulation based on extensions or modifications of Boole's algebra. These efforts met with only limited success. A different approach was taken in 1879 by Frege [60], whose system was the ancestor of the predicate calculus. The latter language has superseded Boolean algebra as a medium for general symbolic reasoning.

The elementary units of discourse in the predicate calculus are the *predicates*, or *atomic formulas*. These are statements such as "X likes Mary," and "$X > Y + 2$," which, for any allowed values of their variables, are either true or false. The variables in a predicate may be *quantified*, by the symbols $\forall$ ("for all") or $\exists$ ("there exists"), to form statements such as $\forall X(X$ likes Mary) or $\exists X(X$ likes Mary); these statements mean, respectively, "everyone likes Mary" and "someone likes Mary." Predicates may be assembled into more complex structures, called *well-formed formulas*, by means of logical connectives such as conjunction (AND), disjunction (OR), and negation (NOT).

Boolean Reasoning

Boolean reasoning builds on the Boole-Schröder algebra of logic, which is based on Boolean equations, rather than on the predicate calculus. Although Boolean equations are predicates—statements that are either true or false for any values of their arguments—almost none of the apparatus of predicate logic is employed in Boolean reasoning. Neither the disjunction of two Boolean equations nor the negation of a Boolean equation is a Boolean equation; thus neither of these operations is generally allowed in Boolean reasoning (see Rudeanu [172, Chapt. 10] for results concerning disjunction and negation of Boolean equations). As shown by Boole, however, the conjunction of two or more Boolean equations is a Boolean equation. The conjunction of a system of equations is therefore expressed by its equivalent single equation, rather than by a symbolic conjunction of equations. Thus the only well-formed formulas of interest in Boolean reasoning are Boolean equations.

Boole and other nineteenth-century logicians based symbolic reasoning on an equation of the *0-normal* form, *i.e.*,

$$f(x_1, \ldots, x_n) = 0 , \qquad (1)$$

derived from, and equivalent to, a given system of Boolean equations (the equivalent *1-normal* form, *i.e.*, $f'(x_1, \ldots, x_n) = 1$, may also be used). A dissertation published in 1937 by A. Blake [10] showed that the consequents of (1) are readily derived from the *prime implicants* of f. The concept of a prime implicant was re-discovered (and named) in 1952 by W.V.O. Quine, who investigated the problem of minimizing the complexity of Boolean formulas. Quine established the theoretical foundations of minimization-theory in a series of papers [161, 162, 163, 164] in the 1950s. The theory of prime implicants has thus arisen independently to serve two quite different ends, *viz.,* Boolean reasoning (Blake) and formula-minimization (Quine).

The approach to Boolean reasoning outlined in this book owes much to Blake's work. Blake's formulation (outlined in Appendix A) anticipates, within the domain of Boolean algebra, the widely-applied resolution principle in predicate logic, given in 1965 by Robinson [168]. Blake's "syllogistic result," for example, corresponds to Robinson's "resolvent."

Boolean Algebra and Switching Theory

Although Boole's algebra did not succeed in expressing, as he had intended, "those operations of the mind by which reasoning is performed" [13, p. 1], it remains in daily use to deal with the simpler mentality of switching circuits. The possibility of applying Boolean algebra to the design of switching systems was first suggested in 1910 by the physicist P. Ehrenfest [54], who proposed in a review of a text by Couturat [41] that Boolean algebra be used in the design of automatic telephone exchanges. Ehrenfest did not, however, supply details as to how it might be done. Papers providing such details appeared independently between 1936 and 1938 in Japan, the United States, and the Soviet Union (it seems that only the results published in the Soviet Union, by Shestakov [185], were based on Ehrenfest's suggestion). The most influential of these papers was Shannon's "Symbolic Analysis of Relay and Switching Circuits" [183], based on his M.S. thesis at the Massachusetts Institute of Technology. Shannon formulated a "calculus of switching circuits," which he showed to be analogous to the calculus of propositions. In a paper published in Japan in 1937, Nakasima [146] identified the same switching calculus with the algebra of sets. Nakasima's paper, the earliest to apply Boolean algebra to switching theory, discussed methods of solving Boolean equations, with the aim of finding an unknown component switching-path when the composite path is known. Nakasima's work seems to have been little noticed in the United States; in the 1950s, however, a number of papers appeared in the U.S. which applied Boolean equation-solving to the design of switching systems [2, 4, 113, 155].

Motivated by problems arising in the design of switching circuits, A. Svoboda [191] proposed construction of a "Boolean Analyzer," a hardware-adjunct to a general-purpose computer, specialized to solve Boolean equations. The applications of such a unit, and of APL programs for solving

Boolean equations, are described in Svoboda and White [193].

Klir and Marin [103] have stated that "The most powerful tool of the modern methodology of switching circuits seems to be the Boolean equations. Their importance for switching theory reminds one of the application of differential equations in electric circuit theory." There remains a curious difference, however, between the way differential equations and Boolean equations are typically applied: Boolean equations are rarely solved. They are manipulated in form but are seldom the basis for systematic reasoning. Contemporary research on Boolean methods in switching tends instead to emphasize formula-minimization. Many writers on applications of Boolean methods believe in fact that the only useful thing to do with Boolean formulas is to simplify them. A widely-used text [84, p. 60] announces that "Almost every problem in Boolean algebra will be found to be some variation of the following statement: 'Given one of the 2^{2^n} functions of n variables, determine from the large number of equivalent expressions of this function one which satisfies some criteria for simplicity.' " Boole would doubtless deem that to be less than full employment for the algebra he designed as an instrument for reasoning.

Although the processes of Boolean reasoning should for practical reasons keep their internal representations relatively simple, minimization is a topic essentially distinct from reasoning. Minimization is therefore not emphasized in this book, notwithstanding its importance both theoretically and in practice. See Brayton, *et al.*, [18] for an an excellent contemporary treatment of minimization and its applications in the design of VLSI (Very Large-Scale Integration) circuits.

An Approach to Boolean Problem-Solving

The central idea in Boolean reasoning, first given by Boole, is to reduce a given system of logical equations to a single equivalent equation of standardized form (*e.g.*, $f = 0$), and then to carry out the desired reasoning on that equation. This preliminary abstraction enables the processes of reasoning to be independent of the form of the original equations.

The primary tactic employed by Boole and later nineteenth-century logicians was to solve $f(x, y, \ldots) = 0$ for certain of its arguments in terms of others. A solution of an equation is a particular kind of antecedent, however, and not necessarily a consequent, of the equation. An important advance was made in 1937 by Blake [10], who showed that the consequents of $f = 0$

are represented economically in the disjunction of the prime implicants of f. We call this disjunction the *Blake canonical form* for f and denote it by $BCF(f)$.

The task of solving a problem based on a collection of Boolean equations may thus be carried out in three major steps:

1. **Reduction.** Condense the equations into a single Boolean equation of the form $f = 0$.

2. **Development.** Construct the Blake canonical form for f, *i.e.,* generate the prime implicants of f.

3. **Reasoning.** Apply a sequence of reasoning-operations, beginning with the Blake form, to solve the problem.

Steps 1 and 2 are independent of the problem to be solved and are readily automated. The sequence of operations to be applied in Step 3, on the other hand, is dependent upon the problem. To employ Boolean reasoning, therefore, the principal task is to select an appropriate sequence of operations to apply to the formula $BCF(f)$. The operational (*i.e.,* functional) basis of Boolean reasoning differentiates it from from the predicate calculus, whose basis is relational. Other differences between the two languages are discussed below.

Boolean Reasoning *vs.* Predicate Logic

The need to incorporate systematic reasoning into the design of switching systems has attracted the attention of engineers to the theorem-proving methods of the predicate calculus. Design based on these methods is summarized by Kabat and Wojcik [95] as follows: "The basic philosophy of the design approach using theorem proving is to represent the elements of the design process as a set of axioms in a formal system (a theory), state the problem of realizability of the target function as a theorem, and prove it in the context of the theory. Once the theorem is proved, an automatic procedure for the recovery of the logic circuit is to be executed to complete the design."

Boolean reasoning differs from the theorem-proving methodology of predicate logic in a number of important ways. Predicates (propositional functions) and propositions are two-valued. Boolean functions, on the other hand, take on values over an underlying set, the *carrier* of the associated

Boolean algebra; the number of elements in the carrier may be 2 or any higher power of 2. The following properties—valid in propositional logic— do not hold in other than two-valued Boolean algebras:

$$F + G = 1 \quad \Longrightarrow \quad F = 1 \quad \text{or} \quad G = 1$$
$$F \neq 1 \quad \Longrightarrow \quad F = 0 \,.$$

Applying propositional calculation-rules to Boolean problems can therefore lead to incorrect results. The denial of a biconditional in propositional logic, for example, can be expressed as a biconditional; thus $\neg(a \longleftrightarrow b)$ is equivalent to $a \longleftrightarrow \neg b$. The denial of a Boolean equation, however, cannot in general be expressed as a Boolean equation; denying the Boolean equation $a = b$ is not the same as asserting the equation $a = b'$.

The principal problem-solving technique in predicate logic is theorem-proving via refutation, *i.e., reductio ad absurdum.* This technique entails the denial of the theorem to be proved. As noted above, however, the denial of a Boolean equation is not a Boolean equation, and thus refutation-based reasoning is not possible in Boolean algebras having other than two values.

Problem-solving in predicate logic entails assigning values to variables over some domain. Any set, *e.g.,* $\{5, John, \clubsuit, cat, \star\}$, may be the domain for a problem in predicate logic. The domain in a Boolean problem, however, is an ordered structure—the carrier of a Boolean algebra. Consequently, the information produced by Boolean reasoning is typically expressed by intervals (cf. Section 2.4.1).

Outline

Chapters 1 through 4 of this book outline the mathematical basis for Boolean reasoning. Chapter 1, included to make the book self-contained, is a brief survey of fundamental concepts such as propositions, predicates, sets, relations, functions and algebraic systems. Chapter 2 treats the classical Boole-Schröder algebra of logic via Huntington's postulates. Several examples of Boolean algebras are discussed, and important theorems are presented; among the latter are the Stone representation theorem, Boole's expansion theorem, and the Löwenheim-Müller verification theorem. Boolean formulas and Boolean functions are defined and the distinction between these two concepts is emphasized. Orthonormal expansions are defined and their utility is examined. The utility of "big" Boolean algebras (those comprising more than two elements) is discussed. Chapter 3 outlines Blake's theory of canonical formulas [10], and the employment of such formulas in deriving and verifying consequents of Boolean equations. Several methods are presented for constructing the Blake canonical form. The Blake form is employed frequently in the remainder of the book; a number of theorems concerning this form, based on Blake's dissertation [10], are given in Appendix A. Chapter 4 introduces the basic operations from which reasoning procedures may be composed. Among such operations are reduction, elimination, expansion, division, and substitution.

Chapters 5 through 7 treat two categories of Boolean reasoning: syllogistic (Chapter 5) and functional (Chapters 6 and 7). Syllogistic reasoning, a direct approach to the solution of problems in propositional logic, is based on constructing a simplified representation of the consequents of the Boolean equation $f = 0$. Functional reasoning, on the other hand, produces functional equations, *i.e.*, statements of the form $x = g(y, z, \ldots)$, related to the equation $f(x, y, z, \ldots) = 0$. Functional antecedents, solutions of $f = 0$, were investigated by Boole and have been the object of much study since; see Rudeanu [172] for an authoritative survey and a complete bibliography. Functional consequents, on the other hand, seem to have received little attention; we discuss the theory of such consequents as well as a number of their applications.

The last two chapters present applications of Boolean reasoning in digital technology. Chapter 8 discusses the identification of a Boolean "black box" by means of an input-output experiment. Chapter 9 concerns multiple-output combinational switching circuits. Emphasis is placed on the problem of specification; the design-problem is formulated as one of solving the specification. A particular class of solutions, which we call *recursive,* corresponds to loop-free circuits which may employ output-signals to help in generating other output-signals.

Proofs are supplied for new results; a proof is given for an established result, however, only if it is particularly instructive. The Boolean calculations entailed in the examples—and those underlying the new results—were carried out using a set of software-tools which the author calls BORIS (Boolean Reasoning In Scheme); these tools are programmed in PC Scheme, a microcomputer-based dialect of Lisp available from Texas Instruments, Inc. BORIS has been invaluable in the exploration and testing of conjectures, building confidence in good conjectures and rudely puncturing bad ones.

Acknowledgments

The author's interest in Boolean methods owes much to the example and assistance of Professor Sergiu Rudeanu of the Faculty of Mathematics, University of Bucharest, Romania. I am indebted to Professor Rudeanu for his careful reading of a preliminary version of this book; full many a logical rock and mathematical shoal was avoided thanks to his comments. I am indebted also to Captain James J. Kainec, U.S. Army, for reading several generations of this book and supplying a large number of helpful comments. Several chapters were read by Dr. Albert W. Small, who made useful suggestions and corrections.

I wish finally to acknowledge a debt beyond measure—to my dear wife, Roberta. Her steadfast support and encouragement made this book possible.

Boolean Reasoning

Chapter 1

Fundamental Concepts

This chapter surveys basic mathematical ideas and language needed in the remainder of the book. The material in this chapter is provided for readers having little experience with the concepts and terminology of modern algebra; other readers may wish to proceed directly to the next chapter. The discussion is informal and only those topics directly applicable to Boolean reasoning are considered. The reader unacquainted with set-theory is cautioned that the sets discussed in this chapter are restricted to be *finite*, *i.e.*, to comprise only a finite number of elements. A text such as that by Halmos [77] should be consulted to gain a balanced understanding of the theory of sets.

1.1   Formulas

If we put a sheet of paper into a typewriter and strike some keys, we produce a *formula* or *expression*. If we strike only the parenthesis-keys, some of the formulas we might type are the following:

$$(\;) \tag{1.1}$$
$$) \, (\;) \tag{1.2}$$
$$(\, (\;) \, (\, (\;) \,) \,) \tag{1.3}$$
$$(\, (\, (\, (\, (\;) \,) \,) \tag{1.4}$$

Formulas may be discussed in terms of two attributes. The first is *syntax*, which is concerned with the way the symbols in a formula are arranged; the second is *semantics*, which is concerned with what the symbols mean.

An important question of syntax is whether a formula in a given class is *well-formed, i.e.,* grammatical or legal according to the rules governing that class. We will discuss the syntax of well-formed parenthesis-strings later in this chapter; our experience with parentheses should tell us, however, that formulas (1.1) and (1.3) are well-formed, whereas (1.2) and (1.4) are not.

Let us venture beyond the parenthesis-keys, to type more elaborate formulas:

$$3 \leq 2 \tag{1.5}$$

$$\text{2 is a prime number.} \tag{1.6}$$

$$\text{140 IF C=40 THEN 120} \tag{1.7}$$

$$\text{140 IF 120 THEN C=40} \tag{1.8}$$

$$\text{QxRPch} \tag{1.9}$$

$$\text{x + + 5 ((} \tag{1.10}$$

$$(\forall x)[x \in \emptyset \implies x \in \{1, 2\}] \tag{1.11}$$

$$\text{There is life beyond our galaxy.} \tag{1.12}$$

$$\text{This statement is false.} \tag{1.13}$$

The formulas above belong to several syntactic classes. The legality of (1.7) and (1.8), for example, must be judged by the rules for BASIC statements, whereas (1.9) is to be judged by the rules of chess-notation [87]. It is nevertheless fairly obvious that (1.8) and (1.10) are not well-formed, and that the other formulas are well-formed.

One isn't much interested in formulas that are not well-formed; from now on, therefore, "formula" will mean "well-formed formula." An important semantic issue concerning such a formula is whether it can be assigned a truth-value (true or false). We now examine that issue.

1.2 Propositions and Predicates

A *proposition* is a formula that is necessarily true or false, but cannot be both. We do not attribute truth or falsehood to parenthesis-strings; hence formulas (1.1) through (1.4) are not propositions. Formulas (1.5), (1.6), and (1.11), however, are propositions; they are false, true, and true, respectively. Formula (1.12) is a proposition, but determining its truth-value requires more information than we now possess. Formula (1.13) is not a proposition: if we assume it to be true, then its content implies that it is false; if we assume it to be false, then its content implies that it is true.

Predicates. The formula

$$x + y \leq 2$$

is not a proposition. It becomes a proposition, however, if any particular pair of numbers is substituted for x and y. Such a formula is called a *predicate* (or propositional form).

More precisely, suppose $P(x_1, \ldots, x_n)$ represents a formula involving variables $x_1, \ldots, x_n$, and suppose that each of the variables can take on values within a certain domain. The domain of x_1 might be certain numbers, that of x_2 might be animals that chew their cud, that of x_3 might be certain propositions, and so on. We say that the formula represented by $P(x_1, \ldots, x_n)$ is an *n-variable predicate* if it becomes a proposition for each allowable substitution of values for $x_1, \ldots, x_n$. A proposition is therefore a 0-variable predicate.

For notational convenience we write $P(X)$ for $P(x_1, \ldots, x_n)$ and we refer to the domains of the variables $x_1, \ldots, x_n$ collectively as the domain of X.

Quantifiers. Given a predicate $P(X)$, where X is assigned values on some domain D, we may wish to announce something about the number of values of X for which $P(X)$ is true. Two statements of this kind, namely

$$\text{For every } X \text{ in } D, P(X) \text{ is true} \tag{1.14}$$

and

$$\text{For at least one } X \text{ in } D, P(X) \text{ is true,} \tag{1.15}$$

are particularly useful. We use the shorthand $(\forall X)$ to mean "for every X in D" (the domain D being understood) and $(\exists X)$ to mean "for at least one X in D." The symbol $\forall$ is called the *universal quantifier;* the symbol $\exists$ is called the *existential quantifier.* Using these quantifiers, we represent formula (1.14) by

$$(\forall X)P(X)$$

and we represent (1.15) by

$$(\exists X)P(X).$$

The formula $P(X)$ is not in general a proposition, because its truth-value may depend on the value assigned to X. The formulas $(\forall X)P(X)$ and $(\exists X)P(X)$, however, are propositions. Suppose, for example, that $P(x, y)$ is the predicate

$$x + y \leq 2$$

and D is the set of real numbers. Then the formula $(\forall x)(\forall y)P(x,y)$ is a false proposition while $(\forall x)(\exists y)P(x,y)$ is a true proposition. The latter form may be read "for all x there exists a y such that $P(x,y)$," indicating a dependence of the possible values of y upon the value of x. This form therefore has a meaning different from that of $(\exists y)(\forall x)P(x,y)$," read "there is a y such that, for all x, $P(x,y)$.

Implication.   A predicate $P(X)$ is said to *imply* a predicate $Q(X)$, written

$$P(X) \Longrightarrow Q(X), \qquad (1.16)$$

in case, for every X in its domain, $Q(X)$ is true if $P(X)$ is true. Put another (and frequently more useful) way, $P(X)$ implies $Q(X)$ if it cannot happen, for any X, that $P(X)$ is true and $Q(X)$ is false. Formula (1.16) is called an *implication*. $P(X)$ is called the *antecedent* of the implication; $Q(X)$ is called the *consequent.*

Here are some examples of implications:

$$\text{x is a real number} \implies x^2 \geq 0 \qquad (1.17)$$
$$\text{There is life beyond our galaxy} \implies 5 \text{ is odd} \qquad (1.18)$$
$$3 \leq 2 \implies \text{It is raining} \qquad (1.19)$$

Formula (1.17) accords with our customary understanding of the word "implies"; there is a logical connection, that is, enabling us to deduce the consequent Q from the antecedent P. In neither (1.18) nor (1.19), however, does such a connection exist. Each of these formulas nevertheless satisfies the definition of an implication. It cannot happen in (1.18) that P is true and Q is false, because Q is true; similarly, it cannot happen in (1.19) that P is true and Q is false, because P is false.

Implications in which the antecedent and consequent have no apparent connection are not as removed from everyday reasoning as one might suppose. We often hear statements of the form, "If the Cubs win the pennant this year, then I'm Sigmund Freud."

Suppose $P(X)$ and $Q(X)$ are predicates. Then each of the following formulas conveys the same information:

$$P(X) \implies Q(X) \qquad (1.20)$$

$$[\text{not } Q(X)] \implies [\text{not } P(X)] \qquad (1.21)$$

$$(\forall X)[[P(X) \text{ and not } Q(X)] \text{ is false}] \qquad (1.22)$$

$$(\forall X)[[[\text{not } P(X)] \text{ or } Q(X)] \text{ is true}] \qquad (1.23)$$

If we wish to prove a theorem of the form (1.20), it may be convenient instead to prove the equivalent statement (1.21). Such an approach is called *proof by contraposition*, or *contrapositive proof.*

Equivalence. Two predicates P(X) and Q(X) are said to be *equivalent*, written

$$P(X) \iff Q(X), \qquad (1.24)$$

provided P(X) and Q(X) are either both true or both false for each X in the domain. Formula (1.24) is called an *equivalence*. The formula

$$x^2 = -1 \iff x = i \text{ or } x = -i,$$

for example, is an equivalence over the domain of complex numbers.

Comparing the definition of equivalence with that of implication, we see that $P(X)$ and $Q(X)$ are equivalent in case the condition

$$[P(X) \implies Q(X)] \quad \text{and} \quad [Q(X) \implies P(X)] \qquad (1.25)$$

is satisfied.

1.3 Sets

A *set* is a collection of objects; the objects in the collection are called its *elements*. Having said that, in order to affirm the view of a set doubtless shared by the reader, we back away and declare the words "set" and "element" to be primitive. They can be described but not defined, being analogous in this respect to the words "line" and "point" in geometry. Our intent therefore is not to say what sets are but to discuss a few of the things that may done with them legally, inasmuch as they provide us with great notational convenience.

An important caution: we consider only *finite sets, i.e.,* sets possessing a finite number of elements, in this book. The assumed finiteness of sets pervades the discussion in this and subsequent chapters, enabling us to think about sets in ways that might not be applicable were the sets infinite.

We write

$$x \in A$$

to signify that x is an element (or member) of a set named A.

Ways of describing sets. A set may be described by *enumeration*, *i.e.*, by an explicit listing of its elements, written within curly braces. Thus the Jones family may be specified by the formula

$$\{\text{Mrs. Jones, Mr. Jones, Emily Jones, Fido}\}.$$

A second way to describe a set is by means of a *membership-property*. To specify a set S in this way, we write

$$S = \{x|P(x)\}$$

where $P(x)$ is a predicate that is true if and only if $x \in S$. The set E of even numbers is thus described by

$$E = \{ \, x \mid x \text{ is a number divisible by 2} \, \}.$$

Some sets are described conveniently either by enumeration or by a membership-property; the choice between the two equivalent specifications

$$S = \{1, -1, i, -i\}$$

and

$$S = \{x|x^4 = 1\},$$

for example, is clearly a matter of taste. The set $\{s, \text{Fido}, \$\}$, on the other hand, would be difficult to describe by means of a membership-property that does not amount to enumeration.

A third way to describe a set is by means of a *recursive rule*. Consider the set

$$B = \{ \begin{bmatrix} 0 \\ 1 \end{bmatrix}, \begin{bmatrix} 0 & 0 \\ 0 & 1 \\ 1 & 0 \\ 1 & 1 \end{bmatrix}, \begin{bmatrix} 0 & 0 & 0 \\ 0 & 0 & 1 \\ 0 & 1 & 0 \\ 0 & 1 & 1 \\ 1 & 0 & 0 \\ 1 & 0 & 1 \\ 1 & 1 & 0 \\ 1 & 1 & 1 \end{bmatrix}, \ldots \}$$

of *binary codes*. This set is described by the following statements:

1. $\begin{bmatrix} 0 \\ 1 \end{bmatrix}$ is an element of B.

2. If $[X]$ is an element of B, then so is $\begin{bmatrix} 0 \\ \vdots \ \ X \\ 0 \\ \hline 1 \\ \vdots \ \ X \\ 1 \end{bmatrix}$.

3. Nothing else is an element of B.

The first statement in the foregoing definition is called the *base* of the definition; it puts certain objects (in this case just one) explicitly in the set. The second statement, called the *recursion*, is a construction-rule, telling us how to manufacture members of the set from other members of the set. The third statement (which is often omitted, being understood) is called the *restriction*; it allows membership in the set only to those objects appointed either by the base or by the recursion. Any of the foregoing parts of a recursive definition may itself have several parts.

Let us consider again the well-formed parenthesis-strings discussed earlier in this chapter. The set of such strings (let us call it P) is described recursively by the statements that follow:

BASE: ( ) is an element of P.
RECURSION: If X and Y are elements of P,
then so are (X) and XY.
RESTRICTION: Nothing else is an element of P.

Suppose we want to show that some property holds for every member of a set S. We can do so, if S is described recursively, in two steps:

Step 1. Show that the property holds for all of the members of S specified by the base.
Step 2. Show that if the property holds for all the members of an arbitrary subset of S, then it must hold for all additional members generated from that subset by recursion.

Abstractness of sets. The elements of a set may be concrete or abstract. A set, however, is always abstract; the only quality a set S possesses is its ability to attach membership-tags to objects. It is important not to confuse properties of a set with properties of the objects it comprises. The set $\{2\}$, for example, is not a number and the set $\{Fido\}$ cannot bark.

Sets vs. sequences. A *sequence* $(x_1, x_2, \ldots, x_n)$ is an ordered collection of objects that are not necessarily distinct. Sequences are sometimes called ordered sets, n-tuples, arrays, or vectors; the delimiters [...] or $< \ldots >$ are sometimes used instead of $(\ldots)$. In a sequence, unlike a set, the order of enumeration of elements is important. Thus $(a, b, c) \neq (b, c, a)$, whereas $\{a, b, c\} = \{b, c, a\}$. Another difference between sets and sequences is that the elements in a set, unlike those in a sequence, must be distinct; thus $\{Fido, \$, Fido, 5\}$ is an illegal representation for a set, whereas $(Fido, \$, Fido, 5)$ is a legal sequence.

Inclusion. If a set S consists entirely of elements that are also members of a set T, then we say that S is *included in* (or is *a subset of*) T and we write

$$S \subseteq T .$$

More formally, we define the relation $\subseteq$ by the equivalence

$$[S \subseteq T] \Longleftrightarrow (\forall x)[x \in S \Longrightarrow x \in T] . \qquad (1.26)$$

Unlike the membership-relation $\in$, which relates elements to sets, the inclusion-relation $\subseteq$ relates sets to sets.

Equality. Two sets S and T are *equal*, written $S = T$, provided each comprises exactly the same elements. Equality of sets is therefore defined by the statement

$$[S = T] \Longleftrightarrow [S \subseteq T \text{ and } T \subseteq S] .$$

We say that S is *properly included* in T in case S is included in T but S is not equal to T.

Cardinality. The number of elements in a finite set S is called the *cardinality* of S and is denoted by $\# S$ (the notations $|S|$, $n(S)$, and $card(S)$ are also used). Thus $\#\{Fido, 1\} = 2$.

The concept of cardinality (or cardinal number) is defined also for infinite sets (see, *e.g.*, Halmos [77]), but clearly entails a generalization of "number of elements." Our concern in this book, however, is exclusively with finite sets.

The empty set. A useful concept is that of the *empty* (or *null*) set, denoted by $\emptyset$. This is the set comprising no elements at all; thus $\emptyset$ is defined by the statement

$$\#\emptyset = 0 \, .$$

The empty set is included in every set, *i.e.,* $\emptyset \subseteq T$ for any set T. To prove that this is the case, based on the definition (1.26) of the relation $\subseteq$, we must show that the proposition

$$(\forall x)[x \in \emptyset \implies x \in T]$$

is true for any set T. The left side of the implication in (1.3), *i.e.,* $x \in \emptyset$, is false for all x; hence the implication is true for all x—and we conclude that $\emptyset$ is a subset of any set. Thus, for example,

$$\{ \, x \mid x \text{ is a flying elephant} \} \subseteq \{\text{Fido},1\}$$

is a valid inclusion.

We speak of "the" empty set because $\emptyset$ is unique (the proof is assigned as homework). Thus

$$\{x \mid x \text{ is a flying elephant}\} = \{x \mid x \text{ is an even integral divisor of 5}\}$$

is a valid statement.

1.4 Operations on Sets

We now discuss some useful ways to make sets out of other sets.

Cartesian product. The *cartesian* (direct, cross) *product* of sets S and T, written $S \times T$, is the set defined by

$$S \times T = \{(x, y) \mid x \in S \text{ and } y \in T\} \, . \tag{1.27}$$

Thus

$$\{a, b\} \times \{a, b, c\} = \{(a, a), (a, b), (a, c), (b, a), (b, b), (b, c)\} \, . \tag{1.28}$$

The cartesian product is not commutative, *i.e.,* $S \times T \neq T \times S$ except for special S, T pairs. The cardinality of $S \times T$ is related to that of the individual sets S and T by

$$\#(S \times T) = (\#S) \cdot (\#T) \, . \tag{1.29}$$

Thus the sets $S \times T$ and $T \times S$ have the same cardinality. In (1.28), for example, $\#S = 2$, $\#T = 3$, and $\#(S \times T) = \#(T \times S) = 6$.

The cartesian product is a set of ordered pairs. We define a three-fold cartesian product as a set of ordered triples, *i.e.*,

$$R \times S \times T = \{(x, y, z) | x \in R, y \in S, z \in T\}. \tag{1.30}$$

Higher-order cartesian products are defined by obvious extension. We write S^n to signify the n-fold cartesian product of S with itself, *i.e.*,

$$S^n = \overbrace{S \times S \times \cdots \times S}^{n \text{ times}}. \tag{1.31}$$

Power set.   The *power set* of a set S, written 2^S, is the set of subsets of S, *i.e.*,

$$2^S = \{R | R \subseteq S\}. \tag{1.32}$$

Thus

$$2^{\{a,b\}} = \{\emptyset, \{a\}, \{b\}, \{a, b\}\}. \tag{1.33}$$

The notation 2^S serves to remind us that the cardinality of 2^S is determined from the cardinality of S by the relation

$$\#(2^S) = 2^{\#S}. \tag{1.34}$$

An alternative notation for the power set is $P(S)$.

Union, intersection, and complement.   Let S and T be sets. Then the *union* of S and T, written $S \cup T$, is defined by

$$S \cup T = \{x | x \in S \text{ or } x \in T \text{ or both}\}. \tag{1.35}$$

Thus $\{a, b, c\} \cup \{a, b, d\} = \{a, b, c, d\}$.

The *intersection* of S and T, written $S \cap T$, is defined by

$$S \cap T = \{x | x \in S \text{ and } x \in T\}. \tag{1.36}$$

Thus $\{a, b, c\} \cap \{a, b, d\} = \{a, b\}$. Two sets S and T are said to be *disjoint* if $S \cap T = \emptyset$, *i.e.*, if they have no elements in common.

The *relative complement* of T with respect to S, written $S - T$ or $S \setminus T$, is defined by

$$S - T = \{x | x \in S \text{ and } x \notin T\}. \tag{1.37}$$

Hachimijiogan
Bee Fit Herbs
4710 Yelm Highway
Lacey, WA 98503
1-888-842-2049
www.1stchineseherbs.com

Eastern Medicine Group
7773 Lake Street
Riverforest, IL 60305
1-708-366-8002
www.kamponinstitute.com

Vincamine
Seacoast Natural Foods
1-800-555-6792
www.seacoastvitamins.com

Smartbomb.com
66 Morris Street
Morristown, NJ 07960
1-800-425-3115
www.smartbomb.com

Huperzine A
Seacoast Natural Foods
1-800-555-6792
www.seacoastvitamins.com

Prohealth
2040 Alameda Padre Serra,
Suite 101
Santa Barbara, CA 93103
1-800-366-5924
www.alzheimerssupport.com

Supplements
Life Extension Foundation
1100 West Commercial
Boulevard
Fort Lauderdale, FL 33309
1-800-544-4440
www.lef.org

Wholesale Nutrition
P.O. Box 3345
Saratoga, CA 95070
1-800-325-2664

Hardbody Nutrition
1-800-378-6787
www.hardbodynutrition.com

of equivalence-classes of an equivalence-relation on S. Thus the equivalence-relations on S are in one-to-one correspondence with the partitions of S. The partition of $\{a, b, c, d, e, f\}$ corresponding to the equivalence-relation (1.43), for example, is

$$\{\{a, b, c\}, \{d, e\}, \{f\}\} = \{[a], [d], [f]\} .$$

The elements a, d, and f in the latter set are called "representatives" of their equivalence-classes; any element of an equivalence-class may be chosen, clearly, to be the representative of its class.

Equivalence is a generalization of equality. We say that two things belonging to a set S are equivalent, even if they aren't equal, if they belong to the same block of a partition of S generated by some classification-scheme. Suppose that S is the set of all people in the world. Then we might say that two people are equivalent if they are of the same sex (thereby partitioning S into two equivalence-classes). Under other schemes of classification, we might say that two people are equivalent if they have the same nationality or are of the same age. As another example, suppose S to be the set of all arithmetic formulas. The elements $x^2 - y^2$ and $(x + y)(x - y)$ of S are clearly not equal (as formulas). They produce the same numerical result, however, if specific numbers are substituted for x and y; hence, we call the two formulas equivalent.

Suppose, as a further example, that S is the set of integers, *i.e.*, $S = \{\ldots, -2, -1, 0, 1, 2, \ldots\}$ and that a relation R is defined on S by the formula

$$(x, y) \in R \Longleftrightarrow x - y \text{ is divisible by 3.} \tag{1.45}$$

The relation R is reflexive, symmetric and transitive (the proof is assigned as homework); hence R is an equivalence-relation. This relation has three equivalence-classes, namely,

$$
\begin{aligned}
[0] &= \{\ldots, -6, -3, 0, 3, 6, \ldots\} \\
[1] &= \{\ldots, -5, -2, 1, 4, 7, \ldots\} \\
[2] &= \{\ldots, -4, -1, 2, 5, 8, \ldots\} .
\end{aligned}
$$

The associated partition of S is $\{[0], [1], [2]\}$.

The typical symbol for an equivalence-relation is $\equiv$; we write $x \equiv y$, that is, in case (x, y) is a member of the set $\equiv$ of ordered pairs. Another way to say this is that $x \equiv y$ if and only if $[x] = [y]$ with respect to the relation

$\equiv$. Referring to the foregoing example, we write $-6 \equiv 3$, $4 \equiv 7$, etc. In this example, the relation $\equiv$ is called *congruence modulo 3*.

Partial-order relations.   Just as an equivalence-relation is a generalization of the relation "equals," a *partial order* is a generalization of the relation "is less than or equal to." We say that a relation R on a set S is a partial order on S in case, for all $x, y, z \in S$, the relation is

(i)	**reflexive:**	$(x, x) \in R$
(ii)	**anti-symmetric:**	$[(x, y) \in R$ and $(y, x) \in R] \Longrightarrow x = y$
(iii)	**transitive:**	$[(x, y) \in R$ and $(y, z) \in R] \Longrightarrow (x, z) \in R$.

The following are some examples of partial-order relations:

- The relation $\leq$ on the set of real numbers is a partial order of a special kind: for any pair x, y of real numbers, either $x \leq y$ or $y \leq x$ holds. Because of this property, $\leq$ is called a *total order* on the set of real numbers.

- If P is the set of partitions on a set S, then the relation of refinement is a partial order on P.

- It S is any set, then the inclusion-relation $\subseteq$ is a partial order on the set 2^S.

- If S is the set $\{1, 2, 3, \ldots\}$ of natural numbers, then the relation "is a divisor of" is a partial order on S.

Partially-ordered sets and Hasse diagrams.   A set S, together with a partial order $\leq$ on S, is an entity called a *partially-ordered set*, designated by the ordered pair $(S, \leq)$. (We use the symbol $\leq$ as a generic representation for a partial-order relation.) In the third of the relations listed above, for example, the pair $(2^S, \subseteq)$ is a partially-ordered set.

A convenient representation for a partially-ordered set $(S, \leq)$ is a *Hasse diagram*. Each element of S is represented as a point in the diagram; the points are connected by lines in such a way that one may trace continuously upward from a point x to another point y if and only if $x \leq y$. To achieve this result with the fewest possible lines, a line is drawn directly upward from x to y if and only if

- $x \leq y$ and
- there is no third point z such that $x \leq z \leq y$.

Figure 1.1 shows Hasse diagrams for several partially-ordered sets.

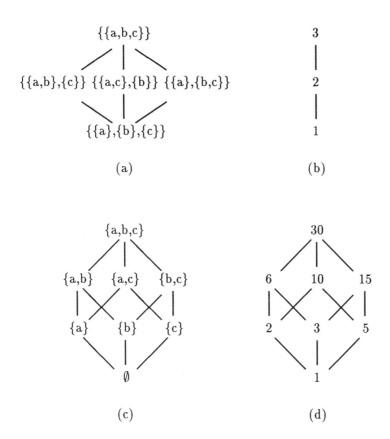

(a) $(\{\pi \mid \pi$ is a partition of $\{a, b, c\}\}$, "is a refinement of")
(b) $(\{1, 2, 3\}$, "less than or equal to")
(c) $(2^{\{a,b,c\}}, \subseteq)$
(d) $(\{1, 2, 3, 5, 6, 10, 15, 30\}$, "is a divisor of")

Figure 1.1: Hasse diagrams.

1.7   Functions

A *function f* from a set S into a set T, written

$$f : S \longrightarrow T , \qquad (1.46)$$

assigns to every element $x \in S$ an element $f(x) \in T$ called the *image* of x. The set S is called the *domain* of f; the set T is called the *co-domain* of f. The *range* of f is the set of images of elements of S under f. The range of f is clearly a subset of its co-domain; if the range of f is equal to T, we say that the function (1.46) is *onto T*.

   A function is a specialized relation. We recall that a relation from S to T is a subset of $S \times T$, *i.e.*, a collection of ordered pairs each of which takes its first element from S and its second element from T. Accordingly, we define a function f from S into T as *a relation from S to T having the property that each element of S appears as a first element in exactly one of the ordered pairs in f.* Thus the formulas "$f(x) = y$" and "$(x, y) \in f$" are equivalent predicates.

Example 1.7.1 Suppose $S = \{a, b, c\}$, $T = \{a, c, d\}$, and define a function f from S into T by the statements

$$
\begin{aligned}
f(a) &= a \\
f(b) &= d \\
f(c) &= a .
\end{aligned}
\qquad (1.47)
$$

Then f is the relation

$$f = \{(a, a), (b, d), (c, a)\} . \qquad (1.48)$$

An alternative way to specify a function—one that is sometimes more convenient than either of the tabulations (1.47) or (1.48)—is by means of a *function-table*. Such a table for the present example is shown in Table 1.1.

n-variable functions.   A function

$$f : S^n \longrightarrow T \qquad (1.49)$$

is called an *n-variable function* from S into T. Thus the temperature-values in a $10' \times 10' \times 10'$ room are described by a 3-variable function from $S = \{x | x$ is a real number between 0 and 10$\}$ into $T = \{x | x$ is a real number$\}$.

x	$f(x)$
a	a
b	d
c	a

Table 1.1: A function-table.

Propositional functions. A function whose co-domain is $\{true, false\}$ is called a *propositional function.* Every predicate, *e.g.,*

$$x + y \le 2 \,, \tag{1.50}$$

represents a propositional function, inasmuch as a predicate becomes a proposition for any allowable substitution of values for its arguments. The formula (1.50) represents a function

$$f : S_1 \times S_2 \longrightarrow \{true, false\} \,,$$

where S_1 and S_2 are sets of numbers. For simplicity, let us take $S_1 = S_2 = \{0, 1, 2\}$. The corresponding propositional function f is a relation comprising nine ordered pairs, namely,

$$
\begin{aligned}
f \quad = \quad & \{((0,0), true), ((0,1), true), ((0,2), true), \\
& ((1,0), true), ((1,1), true), ((1,2), false), \\
& ((2,0), true), ((2,1), false), ((2,2), false)\} \,.
\end{aligned}
$$

Functions vs. formulas. A formula such as

$$f(x) = x^2 - 9$$

is clearly simpler to work with than is the set of ordered pairs defining the function f. The convenience of using formulas to represent functions may lead us to ignore the distinction between these two entities. The distinction is important, however, in the applications of Boolean reasoning. The desired *behavior* of a digital system, for example, is specified by a Boolean function f, whereas the *structure* of the system is specified by an associated Boolean formula. There are typically many equivalent formulas that represent a given function; correspondingly there are typically many circuit-structures that produce a given specified behavior.

1.8   Operations and Algebraic Systems

An *operation*, $\circ$, on a set $S = \{s_1, s_2, \ldots\}$ is a function from $S \times S$ into S, *i.e.*,

$$\circ : S \times S \longrightarrow S . \tag{1.51}$$

To each ordered pair $(a, b) \in S \times S$, the operation $\circ$ assigns an element $a \circ b \in S$. We may specify an operation $\circ$ by an *operation-table* having the following form:

$\circ$	s_1	s_2	$\cdots$
s_1	$s_1 \circ s_1$	$s_1 \circ s_2$	$\cdots$
s_2	$s_2 \circ s_1$	$s_2 \circ s_2$	$\cdots$
$\vdots$			

The pair $(S, \circ)$ is called an *algebraic system*. An example is $([0, 1], \cdot)$, the set of real numbers on the closed interval from 0 to 1, together with the operation of multiplication. An algebraic system may have more than one operation. We are familiar, for example, with the complex field $(K, +, \cdot, 0, 1)$; here K is the set of complex numbers and $+$ and $\cdot$ are addition and multiplication, respectively, of complex numbers. In labelling an algebraic system it is sometimes convenient to name parts of the system in addition to its set and its operations. A *Boolean algebra* $\mathcal{B}$ (to be discussed in Chapter 3), for example, is labelled by the quintuple

$$(B, +, \cdot, 0, 1) . \tag{1.52}$$

The first element in the foregoing specification is a set, the next two elements are operations, and the last two elements are special members of B. The quintuple (1.52) provides five "slots" into which particular sets, operations, and special members may be inserted to form particular Boolean algebras. The algebraic system $(2^S, \cup, \cap, \emptyset, S)$, for example, is the Boolean algebra of subsets of the set S.

Exercises

1. Assuming the domain of x to be the set of real numbers, which of the following is a valid implication?

$$\text{(a)} \quad x^2 = -2 \quad \Longrightarrow \quad x = 5$$
$$\text{(b)} \quad x \leq 3 \quad \Longrightarrow \quad x^2 \geq 0$$
$$\text{(c)} \quad x = 4 \quad \Longrightarrow \quad x \leq 1$$
$$\text{(d)} \quad x + 1 = 2 \quad \Longrightarrow \quad x \leq 3$$

2. Give a recursive definition of the set

$$B = \left\{ \begin{bmatrix} 0 \\ 1 \end{bmatrix}, \begin{bmatrix} 0 & 0 \\ 0 & 1 \\ 1 & 1 \\ 1 & 0 \end{bmatrix}, \begin{bmatrix} 0 & 0 & 0 \\ 0 & 0 & 1 \\ 0 & 1 & 1 \\ 0 & 1 & 0 \\ 1 & 1 & 0 \\ 1 & 1 & 1 \\ 1 & 0 & 1 \\ 1 & 0 & 0 \end{bmatrix}, \ldots \right\}$$

of reflected Gray codes.

3. Let $\emptyset$ denote the empty set.

(a) For an arbitrary set S, which of the following is true?

$$\emptyset \in 2^S$$
$$\emptyset \subseteq 2^S$$

(b) Exhibit the following sets explicitly and state the cardinality of each.

$$2^\emptyset$$
$$2^{\left(2^\emptyset\right)}$$

4. Given $S = \{\emptyset, \{1, 2\}\}$, exhibit 2^S.

5. Let $S = \{\emptyset, 2, \{4,5\}, 4\}$. Decide the truth of each of the following statements:

$$
\begin{array}{lll}
\text{(a)} & \{4,5\} & \subseteq S \\
\text{(b)} & \{4,5\} & \in S \\
\text{(c)} & \{2,4\} & \subseteq S \\
\text{(d)} & \{2,4\} & \in S \\
\text{(e)} & \emptyset & \subseteq S \\
\text{(f)} & \emptyset & \in S \\
\text{(g)} & \{\emptyset\} & \subseteq S \\
\text{(h)} & \{\emptyset\} & \in S \\
\text{(i)} & \{\{4,5\}\} & \subseteq S \\
\text{(j)} & 2 & \subseteq S \\
\text{(k)} & 2 & \in S \\
\text{(l)} & \{2\} & \subseteq S \\
\text{(m)} & \{2\} & \in S
\end{array}
$$

6. Given that S is any non-empty set, decide the truth of each of the following statements. Explain your reasoning in each case.

$$
\begin{array}{lll}
\text{(a)} & S & \in 2^S \\
\text{(b)} & S & \subseteq 2^S \\
\text{(c)} & \{S\} & \in 2^S \\
\text{(d)} & \{S\} & \subseteq 2^S
\end{array}
$$

7. Using the fact that $\emptyset \subseteq T$ for any set T, if $\emptyset$ is an empty set, show that there is only one empty set.

8. Prove or disprove the following statements:

   (a) For all sets S, $S \times \emptyset = \emptyset \times S$.

   (b) If S and T are non-empty sets, then

   $$S \times T = T \times S \quad \Longleftrightarrow \quad S = T .$$

9. How many relations are there from an m-element set to an n-element set?

10. Given sets D and R, define a set F as follows:

$$F = \{f | f : D \longrightarrow R\} .$$

F is the set of functions, that is, that map D into R. Express $\# F$ in terms of $\# D$ and $\# R$.

11. Let S be a set comprising k elements, and let n be a positive integer.

    (a) How many elements are there in S^n?

    (b) How many n-variable functions are there from S into S?

12. Decide, for each of the following sets of ordered pairs, whether the set is a function.

    (a) $\{(x,y) \mid$ x and y are people and x is the mother of y$\}$
    (b) $\{(x,y) \mid$ x and y are people and y is the mother of x$\}$
    (c) $\{(x,y) \mid$ x and y are real numbers and $x^2 + y^2 = 1\}$
    (d) $\{(x,y) \mid [x = 1$ and $y = 2]$ or $[x = -1$ and $y = 2]\}$

Chapter 2

Boolean Algebras

We outline in this chapter the ideas concerning Boolean algebras that we shall need in the remaining chapters. For a formal and complete treatment, see Halmos [78], Mendelson [137], Rudeanu [172], or Sikorski [187]. For an informal approach and a discussion of applications, see Arnold [3], Carvallo [35], Hohn [86], Kuntzmann [110], Svoboda & White [193], or Whitesitt [214]. Rudeanu's text [172] is unique as a complete and modern treatment of Boolean functions and the solution of Boolean equations.

We begin by stating a set of postulates for a Boolean algebra, adapted from those given by Huntington [92].

2.1 Postulates for a Boolean Algebra

Consider a quintuple

$$(\mathbf{B}, +, \cdot, 0, 1) \tag{2.1}$$

in which $\mathbf{B}$ is a set, called the *carrier*; $+$ and $\cdot$ are binary operations on $\mathbf{B}$; and 0 and 1 are distinct members of $\mathbf{B}$. The algebraic system so defined is a *Boolean algebra* provided the following postulates are satisfied:

1. *Commutative Laws.* For all a, b in $\mathbf{B}$,

$$a + b = b + a \tag{2.2}$$

$$a \cdot b = b \cdot a \tag{2.3}$$

2. *Distributive Laws.* For all a, b, c in $\mathbf{B}$,

$$a + (b \cdot c) = (a + b) \cdot (a + c) \tag{2.4}$$

$$a \cdot (b + c) = (a \cdot b) + (a \cdot c) \tag{2.5}$$

23

3. *Identities.* For all a in **B**,

$$0 + a \ = \ a \tag{2.6}$$
$$1 \cdot a \ = \ a \tag{2.7}$$

4. *Complements.* To any element a in **B** there corresponds an element a' in **B** such that

$$a + a' \ = \ 1 \tag{2.8}$$
$$a \cdot a' \ = \ 0 \tag{2.9}$$

(It is readily shown that the element a' is unique.)

We shall be concerned in this book only with *finite* Boolean algebras, *i.e.*, Boolean algebras whose carrier, **B**, is a finite set; thus "Boolean algebra" should be taken invariably to mean "finite Boolean algebra." Although a Boolean algebra is a quintuple, it is customary to speak of "the Boolean algebra **B**," *i.e.*, to refer to a Boolean algebra by its carrier.

As in ordinary algebra, we may omit the symbol "$\cdot$" in forming Boolean products, except where emphasis is desired. Also, we may reduce the number of parentheses in a Boolean expression by assuming that multiplications are performed before additions. Thus the formula $(a \cdot b) + c$ may be expressed more simply as $ab + c$.

2.2   Examples of Boolean Algebras

2.2.1   The Algebra of Classes (Subsets of a Set)

Suppose in a given situation that every set of interest is a subset of a fixed nonempty set S. Then we call S a *universal set* and we call its subsets the *classes* of S. If $S = \{a, b\}$, for example, then the classes of S are $\emptyset$, $\{a\}$, $\{b\}$, and $\{a,b\}$.

The *algebra of classes* consists of the set 2^S (the set of subsets of S), together with two operations on 2^S, namely, $\cup$ (set-union) and $\cap$ (set-intersection). This algebra satisfies the postulates for a Boolean algebra, provided the substitutions

$$
\begin{aligned}
\mathbf{B} &\longleftrightarrow 2^S \\
+ &\longleftrightarrow \cup \\
\cdot &\longleftrightarrow \cap \\
0 &\longleftrightarrow \emptyset \\
1 &\longleftrightarrow S
\end{aligned}
$$

are carried out, *i.e.*, the system

$$(2^S, \cup, \cap, \emptyset, S)$$

is a Boolean algebra. The "algebra of logic" of Boole [13], Carroll [34], Venn [210], and other nineteenth-century logicians was formulated in terms of classes. Carroll's problems, involving classes such as "my poultry," "things able to manage a crocodile," and "persons who are despised," remain popular today as logical puzzles.

2.2.2 The Algebra of Propositional Functions

A proposition is a statement that is necessarily true or false, but which cannot be both. Propositions are elementary units of reasoning; they may be operated upon, and assembled in various patterns, by a system of calculation called the *algebra of propositions*, or the *propositional calculus*. Let P and Q be propositions. The *conjunction* of P and Q, read "P and Q" and symbolized P $\wedge$ Q, is a proposition that is true if and only if both P and Q are true. The *disjunction* of P and Q, read "P or Q" and symbolized by P $\vee$ Q, is a proposition that is false if and only if both P and Q are false. The *negation* of P, read "not P" and symbolized by $\neg$P, is a proposition that is true if P is false and false if P is true. The *conditional with antecedent P and consequent Q*, read "if P then Q" and symbolized by P $\rightarrow$ Q, is defined to have the same truth-value, for all truth-values of P and Q, as the propositional function $\neg$P $\vee$ Q.

Let P be the set of propositional functions of n given variables, let $\square$ be the formula that is always false (contradiction), and let $\blacksquare$ be the formula that is always true (tautology). Then the system

$$(P, \vee, \wedge, \square, \blacksquare)$$

is a Boolean algebra (see Arnold [3] Goodstein [72] or Hohn [86] for a fuller discussion).

2.2.3   Arithmetic Boolean Algebras

Let n be the product of distinct relatively prime numbers, let D_n be the set
of all divisors of n, and let *lcm* and *gcd* denote the operations "least common
multiple" and "greatest common divisor," respectively. Then the system

$$(D_n, lcm, gcd, \underline{1}, n)$$

is a Boolean algebra, a fact first pointed out, apparently, by Bunitskiy [31].
The symbol $\underline{1}$ denotes the integer 1; it is necessary to distinguish the integer
1 from the Boolean 1-element because the integer 1 is the 0-element of an
arithmetic Boolean algebra.

Example 2.2.1 The arithmetic Boolean algebra for $n = 30$ is

$$(\{\underline{1}, 2, 3, 5, 6, 10, 15, 30\}, lcm, gcd, \underline{1}, 30),$$

giving rise to operations such as the following:

$$
\begin{aligned}
6 + 15 &= 30 \\
6 \cdot 15 &= 3 \, .
\end{aligned}
$$

□

2.2.4   The Two-Element Boolean Algebra

The system

$$(\{0, 1\}, +, \cdot, 0, 1)$$

is a Boolean algebra provided $+$ and $\cdot$ are defined by the following operation-
tables:

+	0	1
0	0	1
1	1	1

·	0	1
0	0	0
1	0	1

2.2.5   Summary of Examples

A summary of the foregoing examples of Boolean algebras is given in Table 2.1.

Algebra	B	+	·	0	1
Subsets of S (Classes)	2^S	∪	∩	∅	S
Propositions	n-Variable Propositional Functions	∨	∧	□	■
Arithmetic Boolean Algebra	Divisors of n	lcm	gcd	$\underline{1}$	n
Two-element Boolean Algebra	{0,1}	$\begin{array}{c\|cc} + & 0 & 1 \\ \hline 0 & 0 & 1 \\ 1 & 1 & 1 \end{array}$	$\begin{array}{c\|cc} \cdot & 0 & 1 \\ \hline 0 & 0 & 0 \\ 1 & 0 & 1 \end{array}$	0	1

Table 2.1: Examples of Boolean Algebras.

2.3 The Stone Representation Theorem

The following theorem, first proved by Stone [190], establishes the important result that any finite Boolean algebra (*i.e.,* one whose carrier is of finite size) has the same structure as a class-algebra.

Theorem 2.3.1 *Every finite Boolean algebra is isomorphic to the Boolean algebra of subsets of some finite set S.*

Stone proved that an infinite Boolean algebra is also isomorphic to a set-algebra, though not necessarily to the simple algebra of subsets of a universal set (see Mendelson [137, Chapter 5] or Rosenbloom [170, Chapter 1]).

Some Boolean algebras have exclusive properties, *i.e.,* properties that do not hold for all Boolean algebras. The properties

$$x + y = 1 \quad \text{iff} \quad x = 1 \text{ or } y = 1 \qquad (2.10)$$

$$x \cdot y = 0 \quad \text{iff} \quad x = 0 \text{ or } y = 0 \,, \qquad (2.11)$$

for example, hold only in two-element algebras. The Stone Representation Theorem tells us that finite class-algebras, however, are not specialized; we may reason, with no lack of generality, in terms of the specific and easily visualized concepts of union, intersection, ∅, and S (where S is the "universal set") rather than in terms of the abstract concepts +, ·, 0, and 1. In particular, we are always justified in using the intuitive properties of class-algebras, rather than going back to the postulates, to prove properties valid for all finite Boolean algebras.

2.4   The Inclusion-Relation

We define the relation $\leq$ on a Boolean algebra as follows:

$$a \leq b \text{ if and only if } ab' = 0 . \tag{2.12}$$

This relation is is a partial order, *i.e.*, it is

(a) **reflexive:** $a \leq a$
(b) **antisymmetric:** $a \leq b$ and $b \leq a \Longrightarrow a = b$
(c) **transitive:** $a \leq b$ and $b \leq c \Longrightarrow a \leq c$

A property analogous to (2.12), *viz.*,

$$A \subseteq B \text{ if and only if } A \cap B' = \emptyset ,$$

holds in the algebra of subsets of a set. A and B are arbitrary classes, *i.e.*, subsets of a universal set S. The relation $A \cap B' = \emptyset$ is easily visualized by means of an Euler diagram, as shown in Figure 2.1.

Because the relation $\leq$ in a Boolean algebra **B** corresponds to the relation $\subseteq$ in the subset-algebra isomorphic to **B**, we call $\leq$ the *inclusion-relation*.

It is useful in practice to recognize the equivalence of the following statements:

$$a \leq b \tag{2.13}$$
$$ab' = 0 \tag{2.14}$$
$$a' + b = 1 \tag{2.15}$$
$$b' \leq a' \tag{2.16}$$
$$a + b = b \tag{2.17}$$
$$ab = a . \tag{2.18}$$

The equivalence of (2.13) and (2.14) is announced by definition in (2.12); the equivalence of the remaining pairs is readily verified.

In Table 2.2 we tabulate several relations, defined in specific Boolean algebras, that correspond to the general inclusion-relation $\leq$.

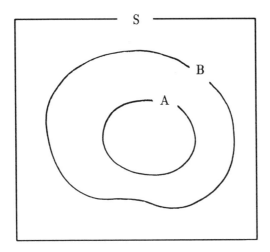

Figure 2.1: Euler diagram illustrating the relation $A \cap B' = \emptyset$.

Boolean Algebra	Relation Corresponding to $\leq$
Subset-Algebra	$\subseteq$
Arithmetic Boolean Algebra	"divides"
Algebra of Propositions	$\longrightarrow$
Two-element Algebra	$\{(0,0),(0,1),(1,1)\}$

Table 2.2: The inclusion-relation in several Boolean algebras.

2.4.1 Intervals

The solutions to many kinds of Boolean problems occur in sets defined by upper and lower bounds. Let a and b be members of a Boolean algebra **B**, and suppose that $a \leq b$. The *interval* (or segment) $[a, b]$ is the set of elements of **B** lying between a and b, *i.e.*,

$$[a, b] = \{x \mid x \in \mathbf{B} \text{ and } a \leq x \leq b\}.$$

Example 2.4.1 Let $\mathbf{B} = \{1, 2, 3, 5, 6, 10, 15, 30\}$ be the 8-element arithmetic Boolean algebra of Example 2.2.1. The inclusion-relation in this algebra is arithmetic divisibility; thus $[3, 30]$ is the interval

$$[3, 30] = \{3, 6, 15, 30\}.$$

$\square$

If $(\mathbf{B}, +, \cdot, 0, 1)$ is a Boolean algebra and a and b are distinct elements of **B** such that $a \leq b$, then the system

$$([a, b], +, \cdot, a, b)$$

is a Boolean algebra.

2.5 Some Useful Properties

We list below some properties—valid for arbitrary elements a, b, c in a Boolean algebra—that are useful in manipulating Boolean expressions.

Property 1 (Associativity):

$$a + (b + c) = (a + b) + c \tag{2.19}$$
$$a(bc) = (ab)c \tag{2.20}$$

Property 2 (Idempotence):

$$a + a = a \tag{2.21}$$
$$aa = a \tag{2.22}$$

Property 3:

$$a + 1 = 1 \tag{2.23}$$
$$a \cdot 0 = 0 \tag{2.24}$$

Property 4 (Absorption):

$$a + (ab) \;=\; a \tag{2.25}$$
$$a(a + b) \;=\; a \tag{2.26}$$

Property 5 (Involution):

$$(a')' \;=\; a \tag{2.27}$$

Property 6 (De Morgan's Laws):

$$(a + b)' \;=\; a'b' \tag{2.28}$$
$$(ab)' \;=\; a' + b' \tag{2.29}$$

Property 7:

$$a + a'b \;=\; a + b \tag{2.30}$$
$$a(a' + b) \;=\; ab \tag{2.31}$$

Property 8 (Consensus):

$$ab + a'c + bc \;=\; ab + a'c \tag{2.32}$$
$$(a + b)(a' + c)(b + c) \;=\; (a + b)(a' + c) \tag{2.33}$$

Property 9:

$$a \;\le\; a + b \tag{2.34}$$
$$ab \;\le\; a \tag{2.35}$$

Property 10 (The principle of duality): Every identity deducible from the postulates of a Boolean algebra is transformed into another identity if

(i)   the operations $+$ and $\cdot$,
(ii) the left and right members of inclusions, and
(iii) the identity-elements 0 and 1

are interchanged throughout.

The postulates themselves, together with the foregoing properties, provide good examples of the duality-principle. Because of that principle, only one of each of the statement-pairs above need be established; the other member of the pair follows by duality.

Proposition 2.5.1 *Let a and b be members of a Boolean algebra. Then*

$$a = 0 \text{ and } b = 0 \quad \text{iff} \quad a + b = 0 \tag{2.36}$$

$$a = 1 \text{ and } b = 1 \quad \text{iff} \quad ab = 1 . \tag{2.37}$$

Proof.   We prove (2.36); the proof of (2.37) follows by dual statements. If $a = 0$ and $b = 0$, then it follows from additive idempotence (2.21) that $a+b = 0+0 = 0$. Conversely, suppose that $a+b = 0$. Multiplying both sides by a and applying distributivity (2.4), multiplicative idempotence (2.22), and property (2.24), we obtain $a + ab = 0$, from which the result $a = 0$ follows by absorption (2.25). We deduce similarly that $b = 0$. $\square$

Proposition 2.5.2 *Let a and b be members of a Boolean algebra. Then*

$$a = b \quad \text{iff} \quad a'b + ab' = 0 . \tag{2.38}$$

Proof.   Suppose $a = b$. Then $a'b + ab' = a'a + aa' = 0 + 0 = 0$. Suppose on the other hand that $a'b + ab' = 0$. If a is added to both sides, the result, after simplification, is $a + b = a$; if instead b is added to both sides, the simplified result is $a + b = b$. Thus $a = b$. $\square$

Proposition 2.5.2 enables an arbitrary Boolean equation to be recast equivalently in the standard form $f = 0$. We will make frequent use of this form. Proposition 2.5.2 also provides a direct means of verifying Boolean identities of the form $a = b$; it is more convenient in many cases to evaluate $a'b + ab'$ than it is to manipulate one side of an identity until it becomes the same as the other.

Example 2.5.1 Let us verify identity (2.30), which has the form $u = v$, where $u = a + a'b$ and $v = a + b$. Thus

$$u'v + uv' = (a + a'b)'(a + b) + (a + a'b)(a + b)' .$$

The latter expression is readily evaluated to be the 0-element; hence, (2.30) is valid. $\square$

Exclusive OR and Exclusive NOR.   The formula $a'b + ab'$ in (2.38) occurs often enough to justify our giving it a special name. It is called the "exclusive OR" (or modulo-2 sum) of a and b and is denoted by $a \oplus b$. This was the sum-operator employed by Boole [13], who denoted it by $+$; the operation we denote by $+$ (the "inclusive OR") was introduced by Jevons [94], who modified Boole's essentially arithmetic algebra to create the more "logical" semantics of modern Boolean algebra. The complement of $a \oplus b$ is called the "exclusive NOR" of a and b, and is denoted by $a \odot b$.

2.6 n-Variable Boolean Formulas

We shall be concerned throughout this book with two kinds of objects: formulas (strings of symbols) and functions. A Boolean function is a mapping that can be described by a Boolean formula; we therefore need to characterize Boolean formulas before discussing Boolean functions.

Given a Boolean algebra **B**, the set of *Boolean formulas* on the n symbols $x_1, x_2, \ldots, x_n$ is defined by the following rules:

1. The elements of **B** are Boolean formulas.

2. The symbols $x_1, x_2, \ldots, x_n$ are Boolean formulas.

3. If g and h are Boolean formulas, then so are

   (a) $(g) + (h)$
   (b) $(g)(h)$
   (c) $(g)'$.

4. A string is a Boolean formula if and only if its being so follows from finitely many applications of rules 1, 2, and 3.

We refer to the strings defined above as *n-variable Boolean formulas*. The number of such formulas is clearly infinite. Given the Boolean algebra $\mathbf{B} = \{0, 1, a', a\}$, the strings

$$
\begin{array}{lll}
\text{a} & & \\
(\text{a}) & + & (\text{a}) \\
(\text{a}) & + & ((\text{a}) + (\text{a})) \\
(\text{a}) & + & ((\text{a}) + ((\text{a}) + (\text{a}))) , \ldots
\end{array}
$$

for example, are all distinct n-variable Boolean formulas for any value of n.

Our definition rejects as Boolean formulas such reasonable-looking strings as $b + x_2$ and ax_1 because they lack the parentheses demanded by our rules. We therefore relax our definition by calling a string a Boolean formula if it can be derived from a Boolean formula by removing, without introducing ambiguity, a parenthesis-pair (...). Thus, the Boolean formula $(b) + (x_2)$ yields, by removal of a parenthesis-pair, the Boolean formula $(b) + x_2$; the latter, by another removal, yields the Boolean formula $b + x_2$.

We are accustomed to thinking of the formulas $(g) + (h)$ and $(g)(h)$ as representing operations (addition and multiplication) in a Boolean algebra.

In the present discussion, however, we are concerned only with syntax, *i.e.,* rules for the formation of strings of symbols. We shall shortly need to view the same formulas as representing functions. In cases where it is necessary to distinguish Boolean formulas from Boolean functions, we indicate that $(g) + (h)$ is a formula by calling it a *disjunction* (rather than a sum) and that $(g)(h)$ is a formula by calling it a *conjunction* (rather than a product). We denote formulas in such situations by upper-case symbols and functions by lower-case. Thus $g + h$ is a function (the sum of functions g and h), whereas $G + H$ is a formula (the disjunction of formulas G and H).

2.7 n-Variable Boolean Functions

An n-variable function $f\colon \mathbf{B}^n \longrightarrow \mathbf{B}$ is called a *Boolean function* if and only if it can be expressed by a Boolean formula. To make this definition precise we must associate a function with each n-variable Boolean formula on $\mathbf{B}$. The set of such formulas is defined recursively by rules given in Section 2.6; we therefore define the n-variable Boolean functions on $\mathbf{B}$ by a parallel set of rules:

1. For any element $b \in \mathbf{B}$, the *constant function*, defined by

$$f(x_1, x_2, \ldots, x_n) = b \qquad \forall (x_1, x_2, \ldots, x_n) \in \mathbf{B}^n \ ,$$

   is an n-variable Boolean function.

2. For any symbol x_i in the set $\{x_1, x_2, \ldots, x_n\}$, the *projection-function*, defined, for any fixed $i \in \{1, 2, \ldots, n\}$, by

$$f(x_1, x_2, \ldots, x_n) = x_i \qquad \forall (x_1, x_2, \ldots, x_n) \in \mathbf{B}^n \ ,$$

   is an n-variable Boolean function.

3. If g and h are n-variable Boolean functions, then the functions $g + h$, gh and g', defined by

   (a) $(g + h)(x_1, x_2, \ldots, x_n) \ = \ g(x_1, x_2, \ldots, x_n) + h(x_1, x_2, \ldots, x_n)$
   (b) $(gh)(x_1, x_2, \ldots, x_n) \ = \ g(x_1, x_2, \ldots, x_n) \cdot h(x_1, x_2, \ldots, x_n)$
   (c) $(g')(x_1, x_2, \ldots, x_n) \ = \ (g(x_1, x_2, \ldots, x_n))' \ ,$

   for all $(x_1, x_2, \ldots, x_n) \in \mathbf{B}^n$, are also n-variable Boolean functions. These functions are said to be defined *pointwise on the range*.

4. Nothing is an n-variable Boolean function unless its being so follows from finitely many applications of rules 1, 2, and 3 above.

Example 2.7.1 Given $\mathbf{B} = \{0, 1, a', a\}$, let us construct the function-table for the two-variable Boolean function $f : \mathbf{B}^2 \longrightarrow \mathbf{B}$ corresponding to the Boolean formula $a'x + ay'$. We observe that the domain,

$$\mathbf{B} \times \mathbf{B} = \{(0,0), (0,1), \ldots, (a,a)\},$$

has 16 elements; hence, the function-table has 16 rows, as shown in Table 2.3. □

Formula			
$a'x + ay'$			

Function – Table

x	y	$f(x, y)$
0	0	a
0	1	0
0	a'	a
0	a	0
1	0	1
1	1	a'
1	a'	1
1	a	a'
a'	0	1
a'	1	a'
a'	a'	1
a'	a	a'
a	0	a
a	1	0
a	a'	a
a	a	0

Table 2.3: Function-table for $a'x + ay'$ over $\{0, 1, a'a\}$.

The rules defining the set of Boolean functions translate each n-variable Boolean formula into a corresponding n-variable Boolean function (which is said to be *represented* by the formula), and every n-variable Boolean function is produced by such a translation.

The number of n-variable Boolean formulas over a finite Boolean algebra **B** is infinite; however, the number of n-variable function-tables over **B**, of which Table 2.3 is an example, is clearly finite. Thus, the relationship between Boolean formulas and Boolean functions is not one-to-one. The distinct Boolean formulas

$$a + x$$
$$ax' + x$$
$$a + a'x$$
$$a + a + x$$
$$(a + x)a + a'x \, ,$$

for example, all represent the same Boolean function.

An important task in many applications of Boolean algebra is to select a good formula—given some definition of "good"—to represent a Boolean function.

The range of a Boolean function. The *range* (*cf.* Section 1.7) of a Boolean function $f\colon \mathbf{B}^n \longrightarrow \mathbf{B}$ is the set of images of elements of $\mathbf{B}^n$ under f. It was shown by Schröder [178, Vol. 1, Sect. 19] that the range of f is the interval

$$[\prod_{A \in \{0,1\}^n} f(A) \, , \, \sum_{A \in \{0,1\}^n} f(A) \,]. \qquad (2.39)$$

A proof for this result is given by Rudeanu [172, Theorem 2.4].

2.8   Boole's Expansion Theorem

The basis for computation with Boolean functions is the expansion theorem given below. Called "the fundamental theorem of Boolean algebra" by Rosenbloom [170], it was discussed in Chapter V of Boole's *Laws of Thought* [13] and was widely applied by Boole and other nineteenth-century logicians. It is frequently attributed to Shannon [184], however, in texts on computer design and switching theory.

Theorem 2.8.1 *If $f\colon \mathbf{B}^n \longrightarrow \mathbf{B}$ is a Boolean function, then*

$$f(x_1, x_2, \ldots, x_n) = x_1' \cdot f(0, x_2, \ldots, x_n) + x_1 \cdot f(1, x_2, \ldots, x_n) \qquad (2.40)$$

for all $(x_1, x_2, \ldots, x_n)$ in $\mathbf{B}^n$.

Proof. We show that (2.40) holds for every n-variable Boolean function. For notational convenience, we write $f(x_1,\ldots)$, $f(0,\ldots)$, and $f(1,\ldots)$, the arguments $x_2,\ldots,x_n$ being understood.

1. Suppose f is a *constant function*, defined by $f(X) = b$ for all $X \in \mathbf{B}^n$, where b is a fixed element of $\mathbf{B}$. Then

$$f(x_1,\ldots) = b = (x_1' + x_1) \cdot b = x_1'b + x_1b = x_1'f(0,\ldots) + x_1f(1,\ldots) ;$$

thus f satisfies (2.40).

2. Suppose f is a *projection-function*, defined by $f(X) = x_i$ for all $X \in \mathbf{B}^n$, where $i \in \{1,\ldots,n\}$. If $i = 1$, then

$$x_1'f(0,\ldots) + x_1f(1,\ldots) = x_1' \cdot 0 + x_1 \cdot 1 = x_1 = f(X) .$$

If $i \neq 1$, then

$$x_1'f(0,\ldots) + x_1f(1,\ldots) = x_1' \cdot x_i + x_1 \cdot x_i = x_i = f(X) .$$

Thus f satisfies (2.40).

3. Suppose (2.40) to hold for n-variable Boolean functions g and h, i.e., suppose the conditions

$$
\begin{aligned}
g(x_1,\ldots) &= x_1'g(0,\ldots) + x_1g(1,\ldots) \\
h(x_1,\ldots) &= x_1'h(0,\ldots) + x_1h(1,\ldots)
\end{aligned}
$$

to be satisfied for all $(x_1, \ldots) \in \mathbf{B}^n$. We show that (2.40) then holds for the n-variable functions $g + h$, gh, and g'.

(a) $(g + h)(x_1, \ldots)$

$\qquad = g(x_1, \ldots) + h(x_1, \ldots)$

$\qquad = (x_1' g(0, \ldots) + x_1 g(1, \ldots)) + (x_1' h(0, \ldots) + x_1 h(1, \ldots))$

$\qquad = x_1'(g(0, \ldots) + h(0, \ldots)) + x_1(g(1, \ldots) + h(1, \ldots))$

$\qquad = x_1'[(g + h)(0, \ldots)] + x_1[(g + h)(1, \ldots)]$.

(b) $(gh)(x_1, \ldots)$

$\qquad = (g(x_1, \ldots))(h(x_1, \ldots))$

$\qquad = (x_1' g(0, \ldots) + x_1 g(1, \ldots)) \cdot (x_1' h(0, \ldots) + x_1 h(1, \ldots))$

$\qquad = x_1'(g(0, \ldots)h(0, \ldots)) + x_1(g(1, \ldots)h(1, \ldots))$

$\qquad = x_1'[(gh)(0, \ldots)] + x_1[(gh)(1, \ldots)]$

(c) $(g')(x_1, \ldots)$

$\qquad = (g(x_1, \ldots))'$

$\qquad = (x_1' g(0, \ldots) + x_1 g(1, \ldots))'$

$\qquad = (x_1 + (g(0, \ldots))')(x_1' + (g(1, \ldots))')$

$\qquad = x_1'[(g')(0, \ldots)] + x_1[(g')(1, \ldots)] + [(g')(0, \ldots)][(g')(1, \ldots)]$

$\qquad = x_1'[(g')(0, \ldots)] + x_1[(g')(1, \ldots)]$.

The final step above is justified by (2.32), the rule of consensus. $\square$

Corollary 2.8.1 *If f is an n-variable Boolean function, then*

$$f(x_1, x_2, \ldots) = [x_1' + f(1, x_2, \ldots)] \cdot [x_1 + f(0, x_2, \ldots)] . \qquad (2.41)$$

If $f \colon \mathbf{B}^n \longrightarrow \mathbf{B}$ is an n-variable Boolean function and if a is an element of $\mathbf{B}$, then the $(n-1)$-variable function $g \colon \mathbf{B}^{n-1} \longrightarrow \mathbf{B}$ defined by

$$g(x_2, \ldots, x_n) = f(a, x_2, \ldots, x_n)$$

is also a Boolean function (the proof is left as an exercise). Thus the functions $f(0, x_2, \ldots, x_n)$ and $f(1, x_2, \ldots, x_n)$ appearing in Theorem 2.8.1 and its corollary are Boolean.

2.9   The Minterm Canonical Form

A Boolean function may be represented by an infinite number of Boolean formulas. It is often useful, however, to work with a restricted class of Boolean formulas, one in which any Boolean function is represented by exactly one formula. A formula in such a class is called a *canonical form*. An important canonical form for Boolean reasoning was given by Blake [10]; we discuss Blake's canonical form in Chapter 3. A canonical form due to Zhegalkin [223], known in the U.S. as the Reed-Muller form [165, 141], is based on the Boolean ring (see Rudeanu [172, Chapt. 1, Sect. 3]). We now consider the *minterm canonical form*, first discussed in Boole's *Laws of Thought* [13, Chapt. V].

Let us develop a 3-variable Boolean function f by repeated application of Boole's expansion theorem (Theorem 2.8.1):

$$
\begin{aligned}
f(x,y,z) &= x'f(0,y,z) + xf(1,y,z) \\
&= x'[y'f(0,0,z) + yf(0,1,z)] + x[y'f(1,0,z) + yf(1,1,z)] \\
&= x'y'z'f(0,0,0) + x'y'zf(0,0,1) \\
&\quad + x'yz'f(0,1,0) + x'yzf(0,1,1) \\
&\quad\quad + xy'z'f(1,0,0) + xy'zf(1,0,1) \\
&\quad\quad + xyz'f(1,1,0) + xyzf(1,1,1) \, .
\end{aligned}
$$

By obvious extension, an arbitrary n-variable Boolean function f may be expanded as

$$
\begin{aligned}
f(x_1,\ldots,x_{n-1},x_n) = \quad & f(0,\ldots,0,0)\, x_1' \cdots x_{n-1}' x_n' \\
+\ & f(0,\ldots,0,1)\, x_1' \cdots x_{n-1}' x_n \\
& \vdots \\
+\ & f(1,\ldots,1,1)\, x_1 \cdots x_{n-1} x_n \, .
\end{aligned}
\tag{2.42}
$$

The values

$$
f(0,\ldots,0,0), \quad f(0,\ldots,0,1), \quad \ldots, \quad f(1,\ldots,1,1)
$$

are elements of **B** called the *discriminants* of the function f; the elementary products

$$
x_1' \cdots x_{n-1}' x_n', \quad x_1' \cdots x_{n-1}' x_n, \quad \ldots, \quad x_1 \cdots x_{n-1} x_n
$$

are called the *minterms* of $X = (x_1,\ldots,x_n)$. (Boole called these products the *constituents* of X.) The discriminants carry all of the information concerning the nature of f; the minterms, which are independent of f, are standardized functional building-blocks. We call the expansion (2.42) the *minterm canonical form* of f and denote it by $MCF(f)$.

For convenience in expressing expansions such as (2.42), we introduce the following notation: for $x \in \mathbf{B}$ and $a \in \{0,1\}$, we define x^a by

$$x^0 = x', \qquad x^1 = x \ . \tag{2.43}$$

This notation is extended to vectors as follows: for $X = (x_1, x_2, \ldots, x_n) \in \mathbf{B}^n$ and $A = (a_1, a_2, \ldots, a_n) \in \{0,1\}^n$, we define X^A by

$$X^A = x_1^{a_1} x_2^{a_2} \cdots x_n^{a_n} \ . \tag{2.44}$$

Let $A = (a_1, a_2, \ldots, a_n)$ and $B = (b_1, b_2, \ldots, b_n)$ be members of $\{0,1\}^n$. Then

$$A^B = \begin{cases} 1 & \text{if } A = B \\ 0 & \text{otherwise.} \end{cases} \tag{2.45}$$

The notation defined above enables us to give the following concise characterization of Boolean functions.

Theorem 2.9.1 *A function $f \colon \mathbf{B}^n \longrightarrow \mathbf{B}$ is Boolean if and only if it can be expressed in the minterm canonical form*

$$f(X) = \sum_{A \in \{0,1\}^n} f(A) X^A \ . \tag{2.46}$$

Proof.   Suppose that f is Boolean. We have deduced by repeated application of Theorem 2.8.1 (Boole's expansion theorem) that f can then be expressed by the minterm form (2.42), which is written equivalently as (2.46). Suppose on the other hand that f can be expressed in the form (2.46). It is clear that (2.46) satisfies the rules given in Section 2.6 for a Boolean formula; thus (2.46) represents a Boolean function. $\square$

Constructing $MCF(f)$ consists in determining the discriminants of f. If f is specified by a function-table, then its discriminants are exhibited explicitly. If f is specified by a Boolean formula, its discriminants may be found by repeated substitutions of 0s and 1s into that formula. Other methods for transforming a Boolean formula into its minterm canonical form are given in texts on logical design and switching theory [106, 122, 136, 218].

Example 2.9.1 Suppose, as in Example 2.7.1, that a two-variable Boolean function $f: \mathbf{B}^2 \longrightarrow \mathbf{B}$ is defined over $\mathbf{B} = \{0, 1, a', a\}$ by the formula $a'x + ay'$. The corresponding 16-row function-table is shown in Table 2.3. The four discriminants of f are obtained from that table as follows:

$$
\begin{aligned}
f(0,0) &= a \\
f(0,1) &= 0 \\
f(1,0) &= 1 \\
f(1,1) &= a' \ .
\end{aligned}
$$

Thus

$$MCF(f) = ax'y' + xy' + a'xy \ . \tag{2.47}$$

□

2.9.1 Truth-tables

If $\mathbf{B}$ has k elements, then the number of rows in the function-table for an n-variable function is $(\#Domain)^n = k^n$. Theorem 2.9.1 implies, however, that *a Boolean function is completely defined by the 0,1 assignments of its arguments*. More precisely, an n-variable Boolean function is defined by the 2^n rows of its function-table for which each argument is either 0 or 1; the sub-table thus specified is called a *truth-table*. The 2-variable function f of Examples 2.7.1 and 2.9.1 is known to be Boolean, inasmuch as it is represented by a Boolean formula. Hence the 4-row truth-table shown in Table 2.4 completely specifies f; the remaining 12 entries in the full function-table (Table 2.3) are determined by the minterm canonical form (2.47), whose coefficients are the entries in Table 2.4.

x	y	$f(x,y)$
0	0	a
0	1	0
1	0	1
1	1	a'

Table 2.4: Truth-table for $a'x + ay'$ over $\{0, 1, a', a\}$.

Example 2.9.2 Given $\mathbf{B} = \{0, 1, a', a\}$, let a 1-variable function f be defined by the function-table shown below. Is f Boolean?

x	$f(x)$
0	a
1	1
a'	a'
a	1

The function f is Boolean if and only if the values of $f(x)$ listed in the table agree, for all $x \in \mathbf{B}$, with the values of $f(x)$ produced by substitution in the minterm canonical form, i.e., $f(x) = x' \cdot f(0) + x \cdot f(1) = x' \cdot a + x \cdot 1$. Substituting the trial-value $x = a$ yields $f(a) = a' \cdot a + a \cdot 1$; thus, $f(a) = a$ if f is a Boolean function. The function-table, however, specifies that $f(a) = 1$; thus f is not a Boolean function. $\square$

2.9.2 Maps

A truth-table for a Boolean function f, e.g., Table 2.4, is a one-dimensional display of the discriminants of f. The same information displayed in two-dimensional form is called a *map, chart* or *diagram*. The display proposed by Karnaugh [99], called a *map*, is widely used by logical designers, and is discussed at length in virtually any text on switching theory or logical design. Karnaugh maps are practical and effective instruments for simplifying Boolean formulas; formula-simplification is a secondary question in Boolean reasoning, however, and a proper discussion of the use of maps would take us afield. We therefore discuss maps only as another way to display the discriminants of a Boolean function. If $\mathbf{B}$ is the two-element algebra $\{0, 1\}$, then the displays we discuss are the ones customarily treated in engineering texts; if $\mathbf{B}$ is larger, the displays are those usually called "variable-entered" [32, 39, 57, 93, 173, 179].

The ancestor of the Karnaugh map was the "logical diagram" proposed by Marquand [131] in 1881. Marquand's diagram was re-discovered by Veitch [208] in 1952; Veitch called it a "chart" and discussed its utility in the design of switching circuits.

Let us develop a Marquand-Veitch diagram for the Boolean function

$$f(x, y, z) = a'z' + ay'z + xz + bx'y' , \tag{2.48}$$

defined over the free Boolean algebra $\mathbf{B} = FB(a, b)$ (*cf.* Section 2.12). In minterm canonical form,

$$f(x, y, z) = \begin{array}{llll} (a' + b)x'y'z' & + & (a + b)x'y'z & + & (a')x'yz' & + & (0)x'yz & + \\ (1)xy'z' & & + & (a)xy'z & & + & (1)xyz' & + & (0)xyz \,. \end{array}$$

The corresponding Marquand-Veitch diagram, shown in Table 2.5, displays the discriminants $f(0, 0, 0), f(0, 0, 1), \ldots, f(1, 1, 1)$ in a 2×4 array according to the natural binary ordering of the arguments in $\{0, 1\}^3$. The discriminants of a four-variable function would be displayed in a 4×4 array, those of a five-variable function in a 4×8 array, and so on. An 8×8 Marquand diagram is shown in Venn's book on symbolic logic [210, p. 140]. Comparing it with his own diagram, Venn said of Marquand's that "there is not the help for the eye here, afforded by keeping all the subdivisions of a single class within one boundary." This visual disadvantage was largely overcome through a modification suggested by Karnaugh in 1953. Karnaugh's map orders the arguments of the discriminants according to the reflected binary code, also called the Gray code [74]. In a Karnaugh map of four or less arguments, the cells for which a given variable is assigned the value 1 form a contiguous band. For maps of more than four variables, not all variables are associated with such bands, but as much help as possible (within the limits of two-dimensional topology) is provided for the eye. A Karnaugh map for the function (2.48) is shown in Table 2.6; the bands for the variables x, y, and z are indicated by lines adjacent to the map.

		\multicolumn{4}{c}{yz}
		00	01	10	11
x	0	$a' + b$	$a + b$	a'	0
	1	1	a	1	0

Table 2.5: Marquand-Veitch diagram for $f(x, y, z) = a'z' + ay'z + xz + bx'y'$.

Marquand, Veitch, and Karnaugh discussed only 0 and 1 as possible cell-entries. The generalization to larger Boolean algebras is immediate, however, if their displays are defined simply as arrays of discriminant-values.

		y		
		z		
	00	01	11	10
0	$a' + b$	$a + b$	0	a'
x	1	1	a	0	1

Table 2.6: Karnaugh map for $f(x, y, z) = a'z' + ay'z + xz + bx'y'$.

2.10   The Löwenheim-Müller Verification Theorem

We define an *identity* in a Boolean algebra **B** to be a statement involving constants (elements of **B**) and arguments $x_1, x_2, \ldots, x_n$ that is valid for all argument-substitutions on $\mathbf{B}^n$ (Boolean identities are discussed further in Section 4.6).

Suppose we wish to verify that an identity, *e.g.*,

$$xy \leq x , \tag{2.49}$$

is valid in all Boolean algebras, and to do so without going back to the postulates. We cannot substitute all possible values for the variables x and y, because no limit has been specified for the size of the carrier, **B**. We have seen, however, that a Boolean function is completely defined by the 0,1 assignments of its arguments. Thus 0,1-substitutions are adequate to verify a Boolean identity. This result, called the *Löwenheim-Müller Verification Theorem* [124, 142], may be stated as follows:

Theorem 2.10.1 *An identity expressed by Boolean formulas is valid in an arbitrary Boolean algebra if and only if it is valid in the two-element Boolean algebra.*

The Verification Theorem applies only to identities, and not to other kinds of properties. Thus the properties (2.10) and (2.11), which are not identities, are valid in the two-element algebra, but not in larger Boolean algebras.

To verify that an identity on n variables is valid for all Boolean algebras, then, it suffices to employ a truth-table, which enables us systematically to make all substitutions on $\{0, 1\}^n$. Table 2.7 illustrates the process for identity (2.49); the identity is valid inasmuch as the asserted relationship (inclusion) holds between xy and x for all argument-substitutions.

x	y	xy	x
0	0	0	0
0	1	0	0
1	0	0	1
1	1	1	1

Table 2.7: Truth-table verifying $xy \leq x$.

2.11 Switching Functions

An n-variable *switching function* is a mapping of the form

$$f: \{0,1\}^n \longrightarrow \{0,1\} . \qquad (2.50)$$

The domain of (2.50) has 2^n elements and the co-domain has 2 elements; hence, there are 2^{2^n} n-variable switching functions.

A truth-table, as we have noted, is a sub-table of a function-table. If **B** is the two-element Boolean algebra $\{0,1\}$, however, a truth-table is identical to the function-table from which it is derived, which leads to the following result.

Proposition 2.11.1 *Every switching function is a Boolean function.*

The term "switching function" may convey the impression that the two-element Boolean algebra suffices to design switching systems. We discuss the utility of larger Boolean algebras for such design in Section 2.18.

Rudeanu [172, p. 17] defines a useful class of functions, the *simple Boolean functions*, that lie between the switching functions and general Boolean functions. A function $f: \mathbf{B}^n \longrightarrow \mathbf{B}$, **B** being an arbitrary Boolean algebra, is a simple Boolean function if f can be represented by a formula built from variables by superposition of the basic operations. Constants, except 0 and 1, are not allowed.

2.12 Incompletely-Specified Boolean Functions

Reasoning in a Boolean algebra **B** frequently entails working with a nonempty interval of Boolean functions, *i.e.*, a set $\mathcal{F}$ defined by

$$\mathcal{F} = \{ f \mid g(X) \leq f(X) \leq h(X) \ \forall X \in \mathbf{B}^n \} , \qquad (2.51)$$

where $g: \mathbf{B}^n \longrightarrow \mathbf{B}$ and $h: \mathbf{B}^n \longrightarrow \mathbf{B}$ are Boolean functions such that $g(X) \leq h(X)$ for all $X \in \mathbf{B}^n$. Inasmuch as g and h are completely defined by their truth-tables, *i.e.*, by their values for argument-assignments on $\{0,1\}^n$, the definition (2.51) for the set $\mathcal{F}$ may be stated equivalently as follows:

$$\mathcal{F} = \{\, f \mid g(X) \leq f(X) \leq h(X) \;\; \forall X \in \{0,1\}^n \,\} \, . \tag{2.52}$$

Let $\mathcal{S}$ be the set of intervals on $\mathbf{B}$; that is,

$$\mathcal{S} = \{\, [a,b] \mid a \in \mathbf{B}, b \in \mathbf{B}, a \leq b \,\} \, . \tag{2.53}$$

Then we may represent the set $\mathcal{F}$ by a single mapping $f: \mathbf{B}^n \longrightarrow \mathcal{S}$, defined as follows:

$$f(X) = [g(X), h(X)] \quad \forall X \in \{0,1\}^n \, . \tag{2.54}$$

A mapping associated in this way with an interval $[g, h]$ is called an *incompletely-specified Boolean function*.

Example 2.12.1 Let $\mathbf{B} = \{0, a', a, 1\}$ and let 2-variable Boolean functions g and h be defined by the formulas

$$\begin{aligned} g(x,y) &= ax' + a'xy \\ h(x,y) &= ax' + a'x + y' \, . \end{aligned} \tag{2.55}$$

Then the 2-variable incompletely-specified function defined by the interval $[g, h]$ is given by the truth-table shown in Table 2.8.

x	y	$f(x,y)$		
0	0	$[a, 1]$	$=$	$\{a, 1\}$
0	1	$[a, a]$	$=$	$\{a\}$
1	0	$[0, 1]$	$=$	$\{0, a', a, 1\}$
1	1	$[a', a']$	$=$	$\{a'\}$

Table 2.8: Truth-table for an incompletely-specified function.

□

Incompletely-specified Boolean functions are most often employed in situations where $\mathbf{B}$ is the two-element Boolean algebra $\{0, 1\}$, in which case g and h are switching functions, and the elements $[0,0], [0,1]$, and $[1,1]$ of $\mathcal{S}$ are renamed 0, X, and 1, respectively. The "value" X (sometimes denoted by the symbol d) indicates a choice between 0 and 1; it is referred to as a "don't-care" value.

Example 2.12.2 Let $B = \{0, 1\}$ and let 2-variable Boolean functions g and h be defined by the formulas

$$
\begin{aligned}
g(x, y) &= x'y + xy' \\
h(x, y) &= x + y .
\end{aligned}
\tag{2.56}
$$

Then the 2-variable incompletely-specified function defined by the interval $[g, h]$ is given by the truth-table shown in Table 2.9.

x	y	$f(x, y)$
0	0	0
0	1	1
1	0	1
1	1	X

Table 2.9: Truth-table for an incompletely-specified switching function.

☐

2.13 Boolean Algebras of Boolean Functions

Let $\mathbf{B}$ be a Boolean algebra comprising k elements and let $F_n(B)$ be the set of n-variable Boolean functions on $\mathbf{B}$. Then the algebraic system

$$(F_n(\mathbf{B}), +, \cdot, 0, 1)$$

is a Boolean algebra in which

+	signifies addition of functions;
·	signifies multiplication of functions;
0	signifies the zero-function; and
1	signifies the one-function.

Example 2.13.1 Suppose $\mathbf{B} = \{0, 1\}$. Then $F_1(\mathbf{B})$ comprises 4 functions, denoted by 0, 1, x', and x. The corresponding truth-tables are shown below.

x	0	1	x'	x
0	0	1	1	0
1	0	1	0	1

☐

2.13.1 Free Boolean Algebras

The elements of $F_n(\mathbf{B})$ are mappings—abstractions for which formulas involving symbols $x_1, x_2, \ldots, x_n$ are merely representations; the symbols themselves are arbitrary. In many applications, however, the symbols $x_1, x_2, \ldots, x_n$ have concrete interpretations. The symbols may in such cases be used to construct a 2^{2^n}-element Boolean algebra as follows: each element of the algebra is the disjunction of a subset of the 2^n minterms built from $x_1, x_2, \ldots, x_n$. The null disjunction is the 0-element of the Boolean algebra; the disjunction of all of the minterms is the 1-element of the algebra. The resulting structure is called the *free Boolean algebra on the n generators* $x_1, x_2, \ldots, x_n$ and is denoted by $FB(x_1, x_2, \ldots, x_n)$. It is shown by Nelson [148, p. 39] that $FB(x_1, x_2, \ldots, x_n)$ is isomorphic to the Boolean algebra of switching functions of n variables.

Example 2.13.2 The carrier of the free Boolean algebra $FB(x_1, x_2)$ is the 16-element set of disjunctions of subsets of the set $\{x_1'x_2', x_1'x_2, x_1x_2', x_1x_2\}$. Each of these formulas is the representative of an equivalence-class of formulas; thus the disjunction $x_1'x_2' + x_1'x_2$ of minterms is equivalent, by the rules of Boolean algebra, to the formula x_1'. $\square$

2.14 Orthonormal Expansions

A set $\{\phi_1, \phi_2, \ldots, \phi_k\}$ of n−variable Boolean functions is called *orthonormal* provided the conditions

$$\phi_i\phi_j = 0 \quad (i \neq j) \tag{2.57}$$

$$\sum_{i=1}^{k} \phi_i = 1 \tag{2.58}$$

are satisfied. A set satisfying (2.57) is called *orthogonal*; a set satisfying (2.58) is called *normal*. An example of an orthonormal set is the set of minterms on $x_1, x_2, \ldots, x_n$.

Given a Boolean function f, an *orthonormal expansion* of f is an expression of the form

$$f(X) = \sum_{i=1}^{k} \alpha_i(X)\phi_i(X), \tag{2.59}$$

where $\alpha_1, \alpha_2, \ldots, \alpha_k$ are n−variable Boolean functions and $\{\phi_1, \ldots, \phi_k\}$ is an orthonormal set. The expansion (2.59) is said to be *with respect to* the set

$\{\phi_1, \ldots, \phi_k\}$. The connection between that set and the coefficients $\alpha_1, \ldots, \alpha_k$ is given by the following result.

Proposition 2.14.1 *Let* $\{\phi_1, \phi_2, \ldots, \phi_k\}$ *be an orthonormal set and let* f *be a Boolean function. Then* f *is given by the expansion (2.59) if and only if*

$$\alpha_i(X)\phi_i(X) = f(X)\phi_i(X) \qquad (\forall i \in \{1, \ldots, k\}) . \tag{2.60}$$

Proof. Suppose the expansion (2.59) to be valid. Then, for any element ϕ_j of $\{\phi_1, \phi_2, \ldots, \phi_k\}$,

$$f(X) \cdot \phi_j(X) = \sum_{i=1}^{k} \alpha_i(X)\phi_i(X)\phi_j(X) .$$

The set $\{\phi_1, \phi_2, \ldots, \phi_k\}$ satisfies the orthogonality-condition (2.57); hence

$$f(X) \cdot \phi_j(X) = \alpha_j(X) \cdot \phi_j(X) ,$$

verifying condition (2.60). Suppose on the other hand that condition (2.60) holds. Then

$$
\begin{aligned}
\sum_{i=1}^{k} \alpha_i(X)\phi_i(X) &= \sum_{i=1}^{k} f(X)\phi_i(X) \\
&= f(X) \cdot \sum_{i=1}^{k} \phi_i(X) \\
&= f(X) ,
\end{aligned}
$$

where we have invoked normality-condition (2.58). $\square$

Using equation-solving techniques discussed in Chapter 6, it can be shown that the set of solutions of equation (2.60) for the coefficient α_i is expressed by

$$\alpha_i \in [f \cdot \phi_i, f + \phi_i'] .$$

An obvious (if uninteresting) solution, therefore, is $\alpha_i = f$. Typically, however, coefficients are arrived at directly by expansion of f, as shown in the following example.

Example 2.14.1 Let a 3-variable Boolean function $f(x, y, z)$ be expressed by the formula $a'xy' + bx'z$ (the Boolean algebra **B** is not specified, but includes the symbols a and b), and suppose we choose the orthonormal set $\{xy', xy, x'\}$. Then an orthonormal expansion of f is

$$f(x, y, z) = a'(xy') + 0(xy) + bz(x') ;$$

that is,

$$
\begin{array}{lcl}
\alpha_1 & = & a' \\
\alpha_2 & = & 0 \\
\alpha_3 & = & bz
\end{array}
\qquad\qquad
\begin{array}{lcl}
\phi_1 & = & xy' \\
\phi_2 & = & xy \\
\phi_3 & = & x' .
\end{array}
$$

An expansion of the same function with respect to the orthonormal set $\{x', x\}$ is

$$f(x, y, z) = (bz)x' + (a'y')x ;$$

that is,

$$
\begin{array}{lcl}
\alpha_1 & = & bz \\
\alpha_2 & = & a'y'
\end{array}
\qquad\qquad
\begin{array}{lcl}
\phi_1 & = & x' \\
\phi_2 & = & x .
\end{array}
$$

□

2.14.1 Löwenheim's Expansions

Suppose that g and h are Boolean functions expressed by expansions with respect to a common orthonormal set $\{\phi_1, \ldots, \phi_k\}$, i.e.,

$$g = \sum_{i=1}^{k} g_i \phi_i \qquad\qquad (2.61)$$

$$h = \sum_{i=1}^{k} h_i \phi_i . \qquad\qquad (2.62)$$

It was shown by Löwenheim [124] that orthonormal expansions of the functions $g + h$, gh, and g' are given by

$$g + h = \sum_{i=1}^{k} (g_i + h_i) \phi_i \qquad\qquad (2.63)$$

$$gh = \sum_{i=1}^{k} (g_i h_i) \phi_i \qquad\qquad (2.64)$$

$$g' = \sum_{i=1}^{k} (g_i') \phi_i . \qquad\qquad (2.65)$$

The foregoing expansions are useful in manipulating Boolean expressions, as the following example illustrates.

Example 2.14.2 Suppose that a Boolean function g is expressed by the formula

$$g = vx'y + wxz' \tag{2.66}$$

and that a formula for g' is desired. A direct application of De Morgan's Laws yields

$$
\begin{aligned}
g' &= (v' + x + y')(w' + x' + z) \\
&= v'w' + v'x' + v'z + w'x + xz + w'y' + x'y' + y'z . \tag{2.67}
\end{aligned}
$$

The complement is easier to calculate, however, and the result has simpler form, if (2.65) is applied. To do so, we seek a simple orthonormal set involving arguments that appear relatively frequently in the formula (2.66). Let us choose the set $\{x', x\}$. An orthonormal expansion of g with respect to this set is

$$g = (vy)x' + (wz')x ,$$

whence, applying (2.65), the complement of g is given by

$$
\begin{aligned}
g' &= (vy)'x' + (wz')'x \\
&= (v' + y')x' + (w' + z)x \\
&= v'x' + x'y' + w'x + xz . \tag{2.68}
\end{aligned}
$$

Formula (2.68) has simpler form, and is obtained more easily (albeit with more planning), than is (2.67). $\square$

Theorem 2.14.1 *Let* $g : \mathbf{B}^n \longrightarrow \mathbf{B}$ *and* $h : \mathbf{B}^n \longrightarrow \mathbf{B}$ *be Boolean functions expressed by expansions with respect to a common orthonormal set* $\{\phi_1, \ldots, \phi_k\}$*, i.e.,*

$$g = \sum_{i=1}^{k} g_i \phi_i$$

$$h = \sum_{i=1}^{k} h_i \phi_i .$$

Let f *be a two-variable Boolean function. Then* $f(g, h)$ *is expressed by the expansion*

$$f(g, h) = \sum_{i=1}^{k} f(g_i, h_i) \phi_i . \tag{2.69}$$

Proof. We begin by applying Theorem 2.8.1 to expand $f(g,h)$:

$$f(g,h) = g'h'f(0,0) + g'hf(0,1) + gh'f(1,0) + ghf(1,1).$$

We then apply the relations (2.63) through (2.65) to produce the further expansion

$$
\begin{aligned}
f(g,h) &= \left(\sum_i g_i'h_i'\phi_i\right)\cdot f(0,0) + \left(\sum_i g_i'h_i\phi_i\right)\cdot f(0,1) + \\
&\quad + \left(\sum_i g_ih_i'\phi_i\right)\cdot f(1,0) + \left(\sum_i g_ih_i\phi_i\right)\cdot f(1,1) \\
&= \sum_i \left[g_i'h_i'f(0,0) + g_i'h_if(0,1) + g_ih_i'f(1,0) + g_ih_if(1,1)\right]\phi_i \\
&= \sum_i f(g_i,h_i)\phi_i\,,
\end{aligned}
$$

which is the desired result. □

Relations (2.63) through (2.65) express three particular "functions of functions" as orthonormal expansions. Theorem 2.14.1 extends these relations to arbitrary two-argument functions; this extension generalizes readily to more than two arguments.

Example 2.14.3 Define f by the formula

$$f = g \oplus h',$$

where g and h are given by

$$
\begin{aligned}
g &= a'x' + bxy' + aby \\
h &= bxy + a'x'.
\end{aligned}
$$

The orthonormal set $\{x', xy', xy\}$, for example, leads to the following expansions of g and h:

$$
\begin{array}{rcccccc}
g &=& (a'+by)x' &+& (b)xy' &+& (ab)xy \\
h &=& (a')x' &+& (0)xy' &+& (b)xy
\end{array}
$$

Thus

$$
\begin{array}{rccccc}
f &=& [(a'+by)\oplus(a')']x' &+& [b\oplus(0)']xy' &+& [ab\oplus(b)']xy \\
&=& [a'(1\oplus0)+a(by\oplus1)]x' &+& [b\oplus1]xy' &+& [b'(0\oplus1)+b(a\oplus0)]xy \\
&=& [a'+a(b'+y')]x' &+& [b']xy' &+& [b'+ba]xy
\end{array}
$$

To simplify computation and avoid error, the coefficients of x' and xy have been been calculated by further expansion, with respect to a and b, respectively. After simplification:

$$f = b' + axy + a'x' + x'y' \ .$$

The utility of orthonormal expansion for hand-computation may be gauged by re-doing the foregoing calculation, applying De Morgan's laws and the definition of Exclusive-OR directly. Orthonormal expansions are advantageous also for machine-computation. $\square$

2.15 Boolean Quotient

Let us define a *letter* in a Boolean formula to be a constant or a variable, and a *literal* to be a letter or its complement. We define a *term* or *product* to be either 1, a single literal, or a conjunction of literals in which no letter appears more than once.

Given a function f and a term t, we define the *quotient of f with respect to t*, denoted by f/t, to be the function formed from f by imposing the constraint $t = 1$ explicitly (f/t is called a *ratio* by Ghazala [70]).

Example 2.15.1 Let a Boolean function f be given by

$$f(w, x, y, z) = w'xz + xy'z' + wx'z \ .$$

The quotient of f with respect to wy' is

$$
\begin{aligned}
f/wy' &= f(1, x, 0, z) \\
&= xz' + x'z.
\end{aligned}
$$

$\square$

It is clear that the function f/t can be represented by a formula that does not involve any variable appearing in the term t. The quotient $f/0$ does not exist, because 0 is not a term (the constraint $0 = 1$, moreover, cannot be satisfied); the quotient $f/1$, on the other hand, is simply f itself, *i.e.*,

$$f/1 = f \ ,$$

because the constraint $1 = 1$ is a satisfied identically. Given terms s and t such that $st \neq 0$, the quotient f/st may be calculated as follows:

$$f/st = (f/s)/t = (f/t)/s \ .$$

Theorem 2.15.1 *Let* $t_1, t_2, \ldots, t_k \colon \mathbf{B}^n \longrightarrow \mathbf{B}$ *be terms and suppose the set* $T = \{t_1, t_2, \ldots, t_k\}$ *to be orthonormal. Let* $f \colon \mathbf{B}^n \longrightarrow \mathbf{B}$ *be a Boolean function. Then* f *is given by the expansions*

$$f(X) = \sum_{i=1}^{k} (f/t_i)(X) \cdot t_i(X) . \qquad (2.70)$$

$$f(X) = \prod_{i=1}^{k} [(f/t_i)(X) + t_i'(X)] . \qquad (2.71)$$

Proof. To prove that expansion (2.70) is valid, we make use of Theorem 2.10.1, the Löwenheim-Müller verification theorem, *i.e.*, we substitute values for X only on $\{0, 1\}^n$. Let $X = A$ be such a substitution. There is exactly one member of T, call it t_j, such that $t_j(A) = 1$ (for $i \neq j$, $t_i(A) = 0$). Thus (2.70) becomes

$$f(A) = (f/t_j)(A) \qquad (t_j(A) = 1)$$

which is a valid identity. To verify expansion (2.71), we use the Löwenheim expansion (2.81) to represent $f'(X)$, beginning with the expansion (2.70), yielding

$$f'(X) = \sum_{i=1}^{k} (f/t_i)'(X) \cdot t_i(X) ,$$

from which (2.71) follows by De Morgan's laws. $\square$

Proposition 2.15.1 *Let* f *and* g *be n-variable Boolean functions and let* t *be an m-variable term* $(m \leq n)$. *Then*

$$f \leq g \quad \Longrightarrow \quad f/t \leq g/t . \qquad (2.72)$$

Proof. The statement $f \leq g$ means that $f(x_1, \ldots, x_n) \leq g(x_1, \ldots, x_n)$ for any choice of the variables $x_1, \ldots, x_n$, in particular for any choice satisfying the constraint $t = 1$. $\square$

Proposition 2.15.2 *Let* f *be a Boolean function and let* t *be a term. Then*

$$f'/t = (f/t)'. \qquad (2.73)$$

Proof.   (By induction on n, the number of arguments appearing in t).
Suppose $n = 1$, *i.e.*, $t = x$, where x is a literal. Then

$$f'(x, y, \ldots)/x = f'(1, y, \ldots) = (f(1, y, \ldots))' = (f/x)' ,$$

where f is evaluated pointwise on the range, *i.e.*, $f'(a, b, \ldots) = (f(a, b, \ldots))'$.
Suppose now that the proposition holds for $n = p$, and let s be a term having
p arguments, whence the augmented term xs has $p + 1$ arguments. Then

$$
\begin{aligned}
f'(x, y, \ldots)/xs &= f'(1, y, \ldots)/s \\
&= (f(1, y, \ldots)/s)' \\
&= (f(x, y, \ldots)/xs)' .
\end{aligned}
$$

Thus the proposition holds for $n = p + 1$. $\square$

Proposition 2.15.3 *Let f be a Boolean function and let t be a term. Then*

$$
\begin{aligned}
t \cdot f &= t \cdot (f/t) & (2.74) \\
t' + f &= t' + (f/t) . & (2.75)
\end{aligned}
$$

Proof.   By Theorem 2.10.1 (the Löwenheim-Müller verification theorem),
identities (2.74) and (2.75) need only be verified for $X \in \{0, 1\}^n$. For any
such value of X, $t(X) \in \{0, 1\}$. Suppose $t(X) = 0$; then (2.74) becomes
the identity $0 = 0$. Suppose on the other hand that $t(X) = 1$; then (2.74)
becomes $f(X) = (f/t)(X)$, which is an identity for $t(X) = 1$ in view of the
definition of (f/t). Identity (2.75) is verified by similar steps. $\square$

Proposition 2.15.4 *Let f be a Boolean function and let t be a term. Then*

$$t \cdot f \le f/t \le t' + f . \qquad (2.76)$$

Proof.   Equation (2.74) is expressed equivalently as $(t \cdot f) \oplus (t \cdot f/t) = 0$,
which is equivalent in turn, after expansion with respect to f/t, to

$$(f/t)'[t \cdot f] + (f/t)[t \cdot f'] = 0 .$$

The latter equation is equivalent to the system

$$
\begin{aligned}
(f/t)'[t \cdot f] &= 0 \\
(f/t)[t \cdot f'] &= 0 ,
\end{aligned}
$$

from which (2.76) follows directly. $\square$

Proposition 2.15.5 *Let p, q, and r be terms such that $pq \neq 0$. Then*

$$pq \leq r \implies q \leq r/p \,.$$

Proof. Form a term $\hat{q}$ from q by deleting any literals in q that are also in p (if every literal in q is also a literal in p, then $\hat{q} = 1$). Then $pq = p\hat{q}$ and

$$pq \leq r \implies p\hat{q} \leq r \tag{2.77}$$
$$\implies p\hat{q}r' = 0 \tag{2.78}$$
$$\implies \hat{q} \leq p' + r \tag{2.79}$$
$$\implies \hat{q} \leq p' + (r/p) \tag{2.80}$$
$$\implies \hat{q} \leq r/p \tag{2.81}$$
$$\implies q \leq r/p \tag{2.82}$$

We invoke Proposition 2.15.3 to produce consequent (2.80). Consequent (2.81) follows from (2.80) because (a) the letters in p, and thus those in p', are distinct from the letters in $\hat{q}$ and (b) the letters in p' are distinct from those in r/p. Finally, (2.82) follows from (2.81) because $q \leq \hat{q}$. $\square$

Proposition 2.15.6 *Let f and g be Boolean functions and let t be a term. Then*

$$g \leq f/t \implies t \cdot g \leq f \,. \tag{2.83}$$

Proof. We evaluate $(t \cdot g) \cdot f'$:

$$\begin{aligned}
(t \cdot g) \cdot f' &= g \cdot (t \cdot f') \\
&= g \cdot t \cdot (f'/t) \quad &\text{(Proposition 2.15.3)} \\
&= g \cdot t \cdot (f/t)' \quad &\text{(Proposition 2.15.2)} \\
&= t \cdot (g \cdot (f/t)')
\end{aligned}$$

If the left member of (2.83) is true then $(g \cdot (f/t)') = 0$; hence $(t \cdot g) \cdot f' = 0$, verifying the right member of (2.83). $\square$

2.16   The Boolean Derivative

Let f be a Boolean function and let x be an argument. We define $\partial f/\partial x$, the *Boolean derivative* of f with respect to x, in terms of the Boolean quotient as follows:

$$\frac{\partial f}{\partial x} = f/x' \oplus f/x \,. \tag{2.84}$$

The concept of such a derivative was introduced by Reed [165] in a discussion of error-correcting codes. Huffman [89] employed the same concept in connection with the solution of Boolean equations and the characterization of information-lossless circuits. Akers [2] called (2.84) the *Boolean difference*. Inasmuch as the term "difference" has another meaning in set-theory and Boolean algebra, we prefer Huffman's term, "derivative." A comprehensive study of Boolean derivatives and their generalizations is to be found the monograph [46] by Davio, Deschamps, and Thayse.

Let us briefly consider an important application of the Boolean derivative, *viz.*, the detection of faults in logical circuits. Suppose that such a circuit has input-signals $x, y, \ldots$ and a single output-signal whose value is specified to be $f(x, y, \ldots)$ for a given function f. Suppose further that the circuit has a *logical fault*, *i.e.*, a condition causing the output to realize a function, g, which differs from f. A *test* for the fault is an input-vector A for which $g(A)$ is different from $f(A)$; thus a vector $(x, y, \ldots)$ is a test for the fault provided it is a solution of the Boolean equation

$$g(x, y, \ldots) = f'(x, y, \ldots) . \qquad (2.85)$$

Many faults arising in practice cause the output to behave as if one of the input-lines, say x, is "stuck" at logical value k (either 0 or 1), so that g is given by

$$g(x, y, \ldots) = f(k, y, \ldots) . \qquad (2.86)$$

In such a case, a test is a solution of the equation

$$f(k, y, \ldots) = f'(x, y, \ldots) , \qquad (2.87)$$

which is equivalent, as we show below, to the system

$$x = k' \qquad (2.88)$$

$$\frac{\partial f}{\partial x} = 1 . \qquad (2.89)$$

Thus a vector $(x, y, \ldots)$ is a test for x stuck-at-k if and only if x satisfies (2.88) and the vector $(y, \ldots)$ satisfies (2.89).

Example 2.16.1 Suppose that a circuit is designed to produce the function

$$f = xy + z .$$

Applying (2.88) and (2.89), a vector (x, y, z) is a test for x stuck-at-k if and only if it is a solution of the system

$$x = k'$$
$$yz' = 1 .$$

Thus a test for x stuck-at-0 is (1,1,0) and a test for x stuck-at-1 is (0,1,0). Tests for y stuck-at-k and z stuck-at-k are similarly derived. $\square$

It remains to show that (2.87) is equivalent to the system (2.88), (2.89). We first observe that (2.87) is equivalent by Proposition 2.5.2 to the equation

$$f(k, y, \ldots) \oplus f'(x, y, \ldots) = 0 . \tag{2.90}$$

Expanding the left side of (2.90) with respect to k and x yields the equation

$$[f(0, y, \ldots) \oplus f(1, y, \ldots)]' + [x \oplus k'] = 0 , \tag{2.91}$$

which is equivalent, in view of definition (2.84), idempotence, and Proposition 2.5.2, to the system composed of (2.88) and (2.89).

The Boolean derivative has been the foundation for research on *Boolean calculus* [11, 202, 203]. The Boolean integral [112, 194, 199], for example, has proven to be a useful concept.

2.17   Recursive Definition of Boolean Functions

Suppose that Boolean functions g and h are expressed by expansions with respect to a common orthonormal set $\{\phi_1, \ldots, \phi_k\}$, *i.e.,*

$$g = \sum_{i=1}^{k} g_i \phi_i$$

$$h = \sum_{i=1}^{k} h_i \phi_i .$$

Let f_1 and f_2 be Boolean functions of one and two variables, respectively. The expansions

$$f_1(g) = \sum_{i=1}^{k} f_1(g_i) \phi_i$$

$$f_2(g, h) = \sum_{i=1}^{k} f_2(g_i, h_i) \phi_i$$

follow from Theorem 2.14.1. If the orthonormal set is $\{x', x\}$, then the foregoing expansions take the form

$$f_1(g) = f_1(g_i/x') \cdot x' + f_1(g_i/x) \cdot x \qquad (2.92)$$
$$f_2(g, h) = f_2(g_i/x', h_i/x') \cdot x' + f_2(g_i/x, h_i/x) \cdot x \qquad (2.93)$$

These expansions (and their obvious extensions to functions of 3, 4, or more variables) are recursive; thus they provide a convenient basis for defining Boolean functions and for calculating with functions of functions.

Let us put the expansions (2.92) and (2.93) in more concrete terms. Suppose F to be a Boolean formula and x to be an argument explicit in F. Assume that the quotient-formula F/x is expressed so as not to involve x explicitly. A recursive definition for a one-variable Boolean function FCN may then be organized as follows:

```
BASE-CASES:     FCN(0)  =  special definition
                FCN(1)  =  special definition
RECURSION:      FCN(F)  =  FCN(F/x') · x' + FCN(F/x) · x
```

The Boolean complement, for example, is defined recursively by the rules

```
COMPLEMENT(0)  =  1
COMPLEMENT(1)  =  0
COMPLEMENT(F)  =  COMPLEMENT(F/x') · x' + COMPLEMENT(F/x) · x
```

Multi-variable functions are similarly defined. The conjunction, CONJ(F,G), for example, is defined by the rules

```
CONJ(0,G)  =  0
CONJ(F,0)  =  0
CONJ(1,G)  =  G
CONJ(F,1)  =  F
CONJ(F,G)  =  CONJ(F/x',G/x') · x' + CONJ(F/x,G/x) · x
```

Such definitions are programmed naturally in non-procedural languages such as Lisp and Prolog, and have been shown by Brayton, *et al.*, [18] to provide an efficient basis for procedural programming.

2.18 What Good are "Big" Boolean Algebras?

We have seen (Section 2.3) that the carrier, $\mathbf{B}$, of a finite Boolean algebra may be any set isomorphic to the set of subsets of some finite set. We have also seen that an n-variable Boolean function f is defined by its *discriminants*, $f(0, \ldots, 0, 0), f(0, \ldots, 0, 1), \ldots, f(1, \ldots, 1, 1)$. Although the discriminants are found by assigning values on $\{0, 1\}^n$ to the argument-vector of f, the *value* of a discriminant may be any element of $\mathbf{B}$; these values are displayed equivalently by a minterm-expansion, a truth-table, or a map.

The specialized two-valued Boolean algebra, $\mathbf{B} = \{0, 1\}$, has properties not shared by its larger cousins; the implication

$$xy = 0 \quad \Longrightarrow \quad x = 0 \text{ or } y = 0 \,,$$

for example, holds only in the two-valued algebra. Thus two-valued thinking does not always translate safely to larger Boolean algebras.

"Big" Boolean algebras (those whose carriers have more than two elements) are needed for the reasoning-techniques discussed in subsequent chapters. In this section, however, we consider the utility of such algebras in everyday applications of Boolean methods, particularly in the design and analysis of switching systems.

The two-valued assumption. The word "Boolean" is often taken in computer science and engineering to mean "two-valued." This interpretation is standard in programming languages; a typical language-manual [14, p. 39] states that "Boolean expressions have one of two possible values: True or False." A term such as "propositional" or "logical" would be better than "Boolean"; however, the expressions defined as Boolean in a procedural programming language are clearly two-valued and are not manipulated as expressions; hence the issue is one simply of terminology. Although the signals in a switching system are also two-valued, the issue in the design of such systems involves more than terminology.

Some writers of texts on logical design define Boolean algebras in a general way but conclude that only the two-valued Boolean algebra is of practical use:

> This algebra is useful for digital switching circuits when [the carrier] is restricted to contain exactly two elements. [61, p. 16].

> The two-valued Boolean algebra (called simply Boolean algebra further on) suffices for our purposes .... [104, p. 67].

> Among all the Boolean algebras, the two-element Boolean algebra **B**$_2$ ..., known as *switching algebra*, is the most useful. It is the mathematical foundation of the analysis and design of switching circuits that make up digital systems. [121, p. 6].

Other—typically older—texts on switching systems [33, 55, 82, 106, 109, 136] work from the outset within an explicitly two-valued switching algebra, following Shannon's [183] propositional formulation of switching theory. We argue however that big algebras play a part in logical design that is both unavoidable (in a certain sense) and useful.

Big algebras can't be avoided. The use of big Boolean algebras in the analysis and design of switching systems is unavoidable, even if unrecognized. Consider for example a digital circuit whose inputs are labelled x, y, and z and whose output, f, is related to its inputs as follows:

$$f = xy + xz' + x'z . \qquad (2.94)$$

One may view equation (2.94) as specifying the value of f in two ways. At any time, the values of the inputs x, y, and z are either 0 or 1; hence, when these values are given, the value of f is determined to be either 0 or 1. An alternative view is that the value of f is a member of the 256-element Boolean algebra of Boolean functions mapping $\{0, 1\}^3$ into $\{0, 1\}$. The latter view of the value of f is necessary in digital design, but is often unconscious. Dietmeyer [50, p. 80] notes in this connection that "many practicing logic designers are unaware that other Boolean algebras exist, even when they use them." [1]

Big algebras are useful. Given a Boolean algebra with carrier **B**, let us recall from Theorem 2.9.1 that a function f mapping **B**^{n} into **B** is Boolean if and only if it can be expressed in the minterm canonical form

$$f(X) = \sum_{A \in \{0,1\}^n} f(A) X^A , \qquad (2.95)$$

where each discriminant, $f(A)$, is an element of **B**. The minterm expansion (2.95) of f, like a truth-table or Karnaugh map, is a way to display the discriminants of f. If **B** $= \{0, 1\}$, therefore, the standard representation-forms

---

[1] The situation is analogous to that of M. Jourdain in Molière's *Le Bourgeois Gentilhomme*, who was astonished to discover that he had been speaking prose for more than forty years.

(expansions, truth-tables, or maps) all display 0s and 1s, fostering the idea that these are the only useful (or even possible) discriminant-values. The practical utility of larger Boolean algebras has nevertheless manifested itself in specialized bendings of the 0-1 assumption. An example is the "variable-entered" Karnaugh map [32, 39, 93, 173, 179], whose entries are allowed to be other than 0 or 1. On p. 157 of his text on digital design, Fletcher [57] writes, "You will find that VEM [Variable-Entered Map] will be one of the most useful design aids discussed and its use permeates the rest of this text in a wide variety of applications." Such maps (and the corresponding minterm-expansions and truth-tables) arise naturally if switching theory is placed on a general Boolean footing.

Example 2.18.1 Let us consider again the Boolean function defined by equation (2.94). Viewed as a three-variable function f_1 over $\mathbf{B} = \{0, 1\}$, this function has the minterm-expansion

$$\begin{aligned} f_1 \;=\; & (0)x'y'z' \;+\; (1)x'y'z \;+\; (0)x'yz' \;+\; (1)x'yz \;+ \\ & (1)xy'z' \;+\; (0)xy'z \;+\; (1)xyz' \;+\; (1)xyz \;. \end{aligned}$$

The corresponding eight-row truth-table is shown on the left in Figure 2.2. An alternative view (one of several possible) is that equation (2.94) defines a two-variable function f_2 over $\mathbf{B} = \{0, 1, x', x\}$; this function has the minterm-expansion

$$f_2 = (x)y'z' + (x')y'z + (x)yz' + (1)yz \;.$$

The four-row truth-table for f_2 is shown on the right in Figure 2.2. $\square$

A Boolean function $f_1(x_1, \ldots, x_m, \ldots, x_n)$ may thus be treated, for $0 \leq m \leq n$, as a Boolean function $f_2(x_1, \ldots, x_m)$ over the free Boolean algebra $FB(x_{m+1}, \ldots, x_n)$.

Example 2.18.2 An n-variable Boolean function f may be realized by a 2^n-to-1 *multiplexer* or *data-selector* [201], which acts as an electronic rotary-switch. A minterm-expansion or truth-table for f translates directly to circuit-connections to the multiplexer. The 2^n discriminants define signals connected to "data" inputs $D_0, D_1, \ldots, D_{2^n-1}$. The n arguments of f define signals connected to "select" inputs $S_{n-1}, \ldots, S_0$; the bit-pattern of the select-inputs defines a number, in binary code, which determines which of the data-inputs is transmitted to the output. Multiplexer-realizations for the

x	y	z	$f_1(x,y,z)$
0	0	0	0
0	0	1	1
0	1	0	0
0	1	1	1
1	0	0	1
1	0	1	0
1	1	0	1
1	1	1	1

$$\mathbf{B} = \{0, 1\}$$

y	z	$f_2(y,z)$
0	0	x
0	1	x'
1	0	x
1	1	1

$$\mathbf{B} = \{0, x', x, 1\}$$

Figure 2.2: Truth-tables for f_1 and f_2.

equivalent functions f_1 and f_2 of Example 2.18.1 are shown in Figure 2.3. If the cost of an inverter (to generate x') plus the cost of the 4-to-1 multiplexer is less than the cost of the 8-to-1 multiplexer, then the "big" Boolean algebra $\{0, x', x, 1\}$ results in a more economical realization than does the two-valued algebra. $\square$

Example 2.18.3 Consider the problem of describing the behavior of a JK flip-flop. The inputs exciting the flip-flop are labelled J and K; the flip-flop's present state is labelled Q. The value of the next state, Q^+, may be expressed in many ways, depending on the choice of the carrier $\mathbf{B}$. The left-hand truth-table in Figure 2.2 expresses Q^+ as a three-variable function over $\mathbf{B} = \{0, 1\}$; the right-hand truth-table expresses Q^+ as a two-variable function over $\mathbf{B} = \{0, Q', Q, 1\}$. The four-row table based on $\mathbf{B} = \{0, Q', Q, 1\}$ expresses the flip-flop's behavior in a more intuitive (and obviously more compact) way than does the eight-row table based on $\mathbf{B} = \{0, 1\}$. $\square$

Conclusion. Variable-assignments other than 0 or 1 (*e.g.*, those to the data-inputs of a multiplexer) are carried out routinely by logical designers. What is sometimes missing is a unified Boolean foundation for such assignments. That foundation is provided by Huntington's postulates and the recognition that (a) the discriminants of a Boolean function may be taken from an arbitrary Boolean algebra, and (b) truth-tables, minterm-expansions and maps serve as equivalent displays of those discriminants.

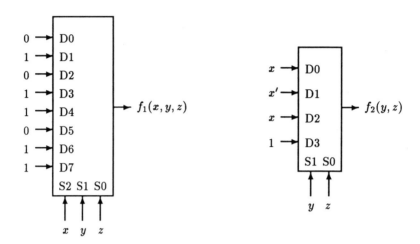

Figure 2.3: Two multiplexer-realizations of $f = xy + xz' + x'z$.

J	K	Q	$Q^+(J,K,Q)$
0	0	0	0
0	0	1	1
0	1	0	0
0	1	1	0
1	0	0	1
1	0	1	1
1	1	0	1
1	1	1	0

$$\mathbf{B} = \{0,1\}$$

J	K	$Q^+(J,K)$
0	0	Q
0	1	0
1	0	1
1	1	Q'

$$\mathbf{B} = \{0, Q', Q, 1\}$$

Figure 2.4: Truth-tables for the JK flip-flop.

Exercises

1. The relation $\leq$ in a Boolean algebra is defined by

$$a \leq b \qquad \text{iff} \qquad ab' = 0 .$$

Prove that the following properties hold for all a, b, c:

(a) $\quad ab \leq a \leq a + c$

(b) $\qquad a = b \qquad \Longleftrightarrow \qquad a \leq b$ and $b \leq a$

(c) $\qquad a \leq b \qquad \Longleftrightarrow \qquad ac \leq bc \quad (\forall c \in \mathbf{B})$

(d) $\qquad a \leq 0 \qquad \Longleftrightarrow \qquad a = 0$

(e) $\qquad 1 \leq a \qquad \Longleftrightarrow \qquad a = 1$

(f) $\qquad a \leq b \qquad \Longrightarrow \qquad \left\{ \begin{array}{ccc} ac & \leq & bc \\ a+c & \leq & b+c \end{array} \right\}$

(g) $\quad \left\{ \begin{array}{ccc} a & \leq & b \\ a & \leq & c \end{array} \right\} \Longrightarrow \quad a \leq bc$

(h) $\quad \left\{ \begin{array}{ccc} a & \leq & c \\ b & \leq & c \end{array} \right\} \Longrightarrow \quad a + b \leq c$

2. Prove or disprove each of the following, assuming that a, b, and c are elements of a Boolean algebra.

(a) $\qquad a + b = a + c \qquad \Longrightarrow \qquad b = c$

(b) $\qquad\qquad ab = ac \qquad \Longrightarrow \qquad b = c$

(c) $\quad \left\{ \begin{array}{ccc} a + b & = & a + c \\ ab & = & ac \end{array} \right\} \Longrightarrow \quad b = c$

3. Given $\mathbf{B} = \{0, 1, a', a\}$, let f be a 2-variable Boolean function for which

$$\begin{aligned} f(0,0) &= 0 \\ f(0,1) &= 1 \\ f(1,0) &= a' \\ f(1,1) &= a . \end{aligned}$$

Find f(a,1).

4. Let $\mathbf{B} = \{0, 1, a', a\}$. Which of the functions f, g, h specified below is a Boolean function?

x	$f(x)$	$g(x)$	$h(x)$
0	a	a'	a'
1	a	1	1
a	a'	1	a'
a'	a'	a'	1

5. Given the 256-element free Boolean algebra $\mathbf{B} = FB(a, b, c)$, let f: $\mathbf{B}^3 \longrightarrow \mathbf{B}$ be a Boolean function for which

$$
\begin{aligned}
f(0,0,0) &= f(0,0,1) &= a \\
f(0,1,0) &= a+b \\
f(0,1,1) &= f(1,0,1) &= f(1,1,1) &= 1 \\
f(1,0,0) &= a+c' \\
f(1,1,0) &= b+c'
\end{aligned}
$$

  (a) Display f by means of a truth-table.

  (b) Write the minterm canonical form for f.

  (c) Write a simplified formula for f.

  (d) Determine $f(a', c, b)$ in as simple a form as you can.

6. Let $\mathbf{B} = \{0, 1, a', a\}$. Find a relation between $f(0)$ and $f(1)$ that is necessary and sufficient for the condition $f(f(x)) = f(x)$ to hold on a Boolean function $f: \mathbf{B}^n \longrightarrow \mathbf{B}$. Make use of the relation to find all Boolean functions satisfying the given condition. Express each such function by a simplified Boolean formula.

7. (McColl [135, 1877-80], cited in Rudeanu [172, Chapter 1]) Show that a function $f: \mathbf{B}^n \longrightarrow \mathbf{B}$ is Boolean if and only if

$$X^A f(X) = X^A f(A) \qquad (\forall A \in \{0,1\}^n).$$

8. Given that f is a Boolean function of one variable, prove the following:

  (a) $f(x+y) + f(xy) = f(x) + f(y)$

  (b) $f(f(0)) = f(0) \cdot f(1) \le f(x) \le f(0) + f(1) = f(f(1))$

  (c) If $x \le y$, then $f(f(x)) \le f(f(y))$.

9. (Poretsky [158], Couturat [41]) Prove the equivalence

$$a \leq x \leq b \quad \Longleftrightarrow \quad x = ax' + bx$$

10. (a) Prove or disprove: for any Boolean function f of one variable, if $x \leq y$, then $f(x) \leq f(y)$.

   (b) If the assertion (a) is not true for all Boolean functions, give a precise characterization of the Boolean functions for which it is true.

11. (a) How many n-variable functions $f: \mathbf{B}^n \longrightarrow B$ (Boolean or not) are there if the set $\mathbf{B}$ has k elements?

   (b) How many of them are Boolean functions?

   (c) How many 3-variable functions are there on $\mathbf{B} = \{0, 1, a', a\}$?

   (d) What fraction of them are Boolean?

12. Let $f: \mathbf{B}^n \longrightarrow B$ be an n-variable Boolean function, and let a be an element of $\mathbf{B}$. Define an $(n-1)$-variable function $g: \mathbf{B}^{n-1} \longrightarrow B$ by the prescription

$$g(x_2, \ldots, x_n) = f(a, x_2, \ldots, x_n) .$$

Show that g is a Boolean function.

13. Let f, g, and h be Boolean functions expressed as

$$f = \sum_{i=1}^{k} f_i \phi_i \qquad g = \sum_{i=1}^{k} g_i \phi_i \qquad h = \sum_{i=1}^{k} h_i \phi_i$$

where the f_i, g_i, and h_i are Boolean functions and $\{\phi_1, \phi_2, \ldots, \phi_k\}$ is an orthonormal set of Boolean functions. Let us consider the expanded functions

$$
\begin{aligned}
f &= & a'(x'y') &+ & b(x'y) &+ & 0(x) \\
g &= & (a+b)(x'y') &+ & 1(x'y) &+ & ab'(x) \\
h &= & ab'(x'y') &+ & a(x'y) &+ & b'(x)
\end{aligned}
$$

for which the orthonormal set is $\{x'y', x'y, x\}$. Suppose that α is a 3-variable Boolean function whose arguments are f, g, and h. Then an orthonormal expansion for α is given as follows in terms of the corresponding expansions for f, g, and h:

$$\alpha(f, g, h) = \sum_{i=1}^{k} \alpha(f_i, g_i, h_i) \cdot \phi_i .$$

Use this expansion-form to calculate the following functions:

(a) f'
(b) $f + g'$
(c) $af + gh$.

14. Use Boole's expansion theorem to prove the following:

$$
\begin{array}{rcl}
\text{(a)} \quad uf(u, v, w) & = & uf(1, v, w) \\
\text{(b)} \quad u'f(u, v, w) & = & u'f(0, v, w) \\
\text{(c)} \quad u + f(u, v, w) & = & u + f(0, v, w) \\
\text{(d)} \quad u' + f(u, v, w) & = & u' + f(1, v, w) \, .
\end{array}
$$

15. If $(\mathbf{B}, +, \cdot, 0, 1)$ is a Boolean algebra and a and b are distinct elements of $\mathbf{B}$ such that $a \le b$, then the system

$$([a, b], +, \cdot, a, b)$$

is a Boolean algebra. Denote by x^* the complement of an element x in this Boolean algebra. Show that

$$x^* = a + bx'$$

where x' is calculated in $\mathbf{B}$.

16. Given

$$
\begin{array}{rcl}
f & = & A'C'D' + A'B'E + A'D'E' + ABC'D' + ACD + B'DE \\
g & = & A'BCE + AC'D + A'C'D' + BC'DE' + ABC'E'
\end{array}
$$

(a) Expand f and g with respect to $\mathbf{B}$ and D, simplifying the discriminants.

(b) Use the expanded forms to calculate

   i. f'
   ii. $f + g'$
   iii. fg
   iv. $f \oplus g$

17. Let **B** be the set $\{0, 1, a', a\}$ and let
$$f: \mathbf{B} \longrightarrow \mathbf{B} \quad \text{and} \quad g: \mathbf{B}^2 \longrightarrow \mathbf{B}$$
be Boolean functions. Given
$$f(0) = a$$
$$f(1) = a'$$
$$g(x, y) = f(x + y) + (f(xy))',$$
express g by a simplified Boolean formula in terms of a, x, and y.

18. For any Boolean algebra **B**, the Boolean functions $f: \mathbf{B} \longrightarrow \mathbf{B}$ in a certain set satisfy the identity
$$f(f(x)) = (f(x))'$$
for all elements $x \in \mathbf{B}$. Describe this set of functions in simple terms. Explain your method clearly.

19. Let $\mathbf{B} = \{0, 1, a', a\}$. List the Boolean functions $f: \mathbf{B} \longrightarrow \mathbf{B}$ that satisfy both of the conditions

(a) $f(f(x)) = f(x)$ $\quad (\forall x \in \mathbf{B})$
(b) $f(0) = a'$.

Express each such function by a simplified Boolean formula. Employ systematic reasoning rather than exhaustive trials. Explain your method clearly.

20. Let **B** be the set $\{0, 1, a', a\}$. How many Boolean functions $f: \mathbf{B}^2 \longrightarrow \mathbf{B}$ are there that satisfy the condition
$$xy \le f(x, y) \le x' + y$$
for all $(x, y) \in \mathbf{B}^2$? Do not determine the functions explicitly.

21. Given the set $\mathbf{B} = \{0, 1, a', a\}$, how many two-variable functions $f: \mathbf{B}^2 \longrightarrow \mathbf{B}$ are there? How many of these are Boolean functions?

22. Let g and h be single-variable Boolean functions. For each of the following cases, express $f(0)$ and $f(1)$ as simplified formulas involving $g(0)$, $g(1)$, $h(0)$, and $h(1)$.

(a) $f(x) = g(h(x))$
(b) $f(x) = g(g'(x))$

Chapter 3

The Blake Canonical Form

Boole's object in inventing an "algebra of logic" [12, 13] was to reduce the processes of reasoning to those of calculation. He showed that a system of logical equations, unlike a system of ordinary equations, can be reduced to a single equivalent equation (we consider the reduction-process in Chapter 4). He chose the standard reduced form $f = 0$, where f is a Boolean function. Reasoning is carried out in Boole's formulation by solving that equation for certain of its arguments in terms of others.

In spite of the efforts of a number of nineteenth-century logicians to extend and generalize Boole's algebra, it proved incapable of representing ordinary discourse. Modern symbolic logic is therefore based on a different system, the predicate calculus, which grew out of the work of Frege [60]. Boole's approach is specialized, moreover, even within the domain of Boolean problems. His system of reasoning, *i.e.*, equation-solving, does not produce logical consequents of $f = 0$; instead, it produces specialized logical antecedents. Boolean equation-solving, which we discuss in Chapter 6, nevertheless has many useful applications,

The first general treatment of both antecedent and consequent Boolean reasoning was that of Poretsky [159] in 1898. Poretsky's system was based on exhaustive tables of antecedents and consequents. The growth of these tables with the number of variables is so rapid, however, as to make them useless for applications.

A more practical approach to Boolean reasoning was developed by Blake [10] in 1937. Blake showed that all of the consequents of $f = 0$, *i.e.*, all Boolean equations $g = 0$ such that the implication

$$f = 0 \quad \Longrightarrow \quad g = 0$$

holds, can easily be generated if f is expressed in a certain canonical form. This form turns out to be the disjunction of all of the prime implicants of f. The term "prime implicant," as well as the theory of systematic formula-minimization in terms of prime implicants, comes from a series of papers by V. W. Quine [161, 162, 163, 164] published between 1952 and 1959. Quine demonstrated that a simplified sum-of-products (SOP) formula for f is necessarily a disjunction of prime implicants of f. He also presented methods for generating all of the prime implicants of f and gave a tabular procedure for selecting a subset of the prime implicants whose disjunction is a simplified SOP formula for f.

While Quine's objective was *Boolean minimization, i.e.,* the discovery of simplified formulas representing a given Boolean function f, Blake's objective was *Boolean inference, i.e.,* the extraction of conclusions from a collection of Boolean data. The theory of prime implicants has thus arisen independently for two quite different applications.

The approach to Boolean inference developed in this book is grounded in Blake's theory of *syllogistic formulas*. We outline that theory in Appendix A, which is adapted from Chapter II of Blake's dissertation. We discuss reasoning based on syllogistic formulas in Chapter 5; our object in this chapter is to define Blake's canonical form (a specialized syllogistic formula) and to describe several methods for its construction.

3.1 Definitions and Terminology

The concept of a Boolean formula on the symbols $x_1, x_2, \ldots, x_n$ was defined in Section 2.6, principally as a stepping-stone to the concept of a Boolean function on the same symbols. Our investigation now focuses on Boolean formulas; hence further definitions are required.

A Boolean function $f: B^n \longrightarrow B$ may be expressed by a variety of formulas. These are built up from *letters, i.e.,* constants $a_1, a_2, \ldots, a_k$ (elements of B) and *variables* $x_1, x_2, \ldots, x_n$, together with the notations of complementation, conjunction, and disjunction. A *literal* is a letter or its complement. A *term* or *product* is either 1, a single literal, or a conjunction of literals in which no letter appears more than once; an *alterm* is either 0, a single literal, or a disjunction of literals in which no letter appears more than once. A *sum-of-products (SOP) formula* is either 0, a single term, or a disjunction of terms; a *product of sums (POS) formula* is either 1, a single alterm, or a conjunction of alterms.

We assume in this chapter, unless stated otherwise, that Boolean functions are expressed by SOP formulas. A Boolean function will be denoted by a lower-case letter (*e.g.*, *f*) and an SOP formula expressing that function by the corresponding upper-case letter (*F*). Terms, viewed either as formulas or functions, will be represented by the lower-case letters $p, q, r, \ldots$. Given an *m*-term SOP formula *F* and an *n*-term SOP formula *G*, $F + G$ is an $(m + n)$-term formula containing all of the terms of *F* together with all those of *G*.

Two SOP formulas will be called *equivalent* ($\equiv$) in case they represent the same Boolean function, *i.e.*, in case one can be transformed into the other, in a finite number of steps, by application of the rules of Boolean algebra. Following Blake, we call two SOP formulas *congruent* ($\overset{\circ}{=}$) in case one can be transformed into the other using only the commutative rule. Thus congruent SOP formulas may differ only in the order of enumeration of their terms and in the order of the literals in any term.

Given two Boolean functions *g* and *h*, we say that *g* is *included in h*, written $g \leq h$, in case the identity $gh' = 0$ is satisfied. When applied to formulas (*e.g.*, $G \leq H$), the relation $\leq$ is inherited from the functions those formulas represent.

An *implicant* of a Boolean function *f* is a term *p* such that $p \leq f$. Any term of an SOP formula for *f* is clearly an implicant of *f*. A *prime implicant* of *f* is an implicant of *f* that ceases to be so if any of its literals is removed. It is shown in Appendix A that an implicant *p* of *f* is a prime implicant of *f* in case, for any term *q*,

$$p \leq q \leq f \implies p = q .$$

An SOP formula *F* will be called *absorptive* in case no term in *F* is absorbed by any other term in *F*. If *F* is not absorptive, then an equivalent absorptive formula, which we call $ABS(F)$, may be obtained from *F* by successive deletion of terms absorbed by other terms in *F*. It is shown in Appendix A that, for any SOP formula *F*, the formula $ABS(F)$ is unique to within congruence.

3.2   Syllogistic & Blake Canonical Formulas

Let *F* and *G* be SOP formulas. We say that *G* is *formally included* in *F*, written $G \ll F$, in case each term of *G* is included in some term of *F*. We write $G \not\ll F$ if *G* is not formally included in *F*. Formal inclusion clearly implies inclusion, *i.e.*, $G \ll F \Longrightarrow G \leq F$ for any F, G pair. The converse does not hold, however, as the following example illustrates.

Example 3.2.1 Let SOP formulas F_1, G, and H be defined as follows:

$$F_1 \;=\; wy' + w'z + w'x'y + wx'yz' \tag{3.1}$$
$$G \;=\; w'y'z + w'x'y \tag{3.2}$$
$$H \;=\; xy'z + x'yz' \,. \tag{3.3}$$

The relations $G \leq F_1$ and $H \leq F_1$ hold for the foregoing formulas. Each term of G is included in a term of F_1; hence $G \ll F_1$. Such is not the case, however, for H, *i.e.*, $H \nll F_1$.

A Formula F will be called *syllogistic* in case, for every SOP formula G,

$$G \leq F \Longrightarrow G \ll F \,.$$

Thus F is syllogistic if and only if every implicant of F is included in some term of F.

Example 3.2.2 The formula

$$F_2 = wy' + w'z + w'x'y + x'yz' + y'z + wx'z' \tag{3.4}$$

is a syllogistic formula equivalent to the formula F_1 in Example 3.2.1 (we discuss the construction of such formulas in the remainder of this chapter). Every SOP formula included in F_2 (or, equivalently, included in F_1) is therefore formally included in F_2. In particular, as the reader should verify, the formulas G and H in Example 3.2.1 are formally included in F_2.

Given SOP formulas F and G, we define $F \times G$ to be the SOP formula produced by multiplying out the conjunction FG, using the distributive laws. If $F = \sum_i s_i$ and $G = \sum_j t_j$, then

$$F \times G = \sum_i \sum_j s_i \cdot t_j \,,$$

where repeated literals are dropped in each product $s_i \cdot t_j$ of terms, $s_i \cdot 1 = s_i$, and $1 \cdot t_j = 1$. A product is dropped if it contains a complementary pair of literals. Thus, for example,

$$(x'y + xz) \times (wx + y + z) = x'y + x'yz + wxz + xyz + xz \,.$$

Let a be any letter. Two terms will be said to have an *opposition* in case one term contains the literal a and the other the literal a'. (If the symbol x stands for the literal a', then we shall understand x' to stand for a.) The terms $x'yz$ and $wy'z$, for example, have a single opposition, in the letter y. Suppose two terms r and s have exactly one opposition. Then the *consensus* [161] of r and s, which we shall denote by $c(r, s)$, is the term obtained from the conjunction rs by deleting the two opposed literals as well as any repeated literals. Thus $c(x'yz, wy'z) = wx'z$. The consensus $c(r, s)$ does not exist if the number of oppositions between r and s is other than one. The consensus of two terms was called their "syllogistic result" by Blake.

Let F be a syllogistic formula for a Boolean function f. We call the formula ABS(F) the *Blake canonical form* for f, and we denote it by $BCF(f)$. Blake showed that $BCF(f)$ is minimal within the class of syllogistic formulas for f, *i.e.*, the set of terms in any syllogistic formula for f is a superset of the set of terms in $BCF(f)$.

The following results are proved in Appendix A:

1. If formulas $F_1, F_2, \ldots, F_k$ are syllogistic, then the formula $F_1 \times F_2 \times \cdots \times F_k$ is also syllogistic.

2. If an SOP formula F is not syllogistic, it contains terms p and q, having exactly one opposition, such that $c(p, q)$ is not formally included in F.

3. Let F be an SOP formula for a Boolean function f. Then F is syllogistic if and only if every prime implicant of f is a term of F.

4. $BCF(f)$ is the disjunction of all of the prime implicants of f.

3.3   Generation of $BCF(f)$

Quine's minimization theory [161, 162, 163, 164] has stimulated a large body of research concerning the efficient generation of $BCF(f)$. We outline in this section the principal approaches; for more details see the survey by Reusch and Detering [167].

$BCF(f)$ is defined to be $ABS(F)$, where F is a syllogistic formula for f; therefore $BCF(f)$ may be generated by the following two-step procedure:

> Step 1.   **Find a syllogistic formula for f.**
> Step 2.   **Delete absorbed terms.**

Blake discussed three approaches for carrying out Step 1; we categorize these as *exhaustion of implicants*, *iterated consensus*, and *multiplication*. Each of the techniques in the extensive literature on the generation of prime implicants appears to belong to one of these categories.

3.4   Exhaustion of Implicants

An obviously syllogistic formula for f is the disjunction of all of the implicants of f (*i.e.*, all terms t such that $t \leq f$). This disjunction was called the "complete canonical form" by Blake. A number of special-purpose logical computers have been designed to perform Step 1 of the foregoing two-step procedure by generating the complete canonical form. These machines employ an n-digit ternary counter, each digit of which corresponds to a letter. The three values of a digit correspond to the ways in which a letter may appear in a term: uncomplemented, complemented, or not at all. The machines designed by Svoboda [191, 192] (USA) and Florine [58, 59] (Belgium) generate all possible terms on the given letters; the machine of Gomez-Gonzalez [73] (Spain) stops generating terms when certain conditions are met. Step 2 (absorption) is typically incorporated by such machines into the process of generating, testing, and storing terms as follows: the one-letter terms are generated first, then the two-letter terms, and so on. A given term is stored if and only if it is an implicant of f and is not an implicant of a term already stored.

Although the foregoing method is simple in conception and readily programmed (or implemented in hardware) for low values of n, the number of candidate-terms that must be stored, and the time required for the requisite scanning, rises exponentially with n. The number of terms on n letters is $3^n - 1$; for 20 letters, this number is about 3.5 billion. Also, many terms are likely to be annihilated in Step 2 of the two-step procedure discussed above. Much effort, beginning in the mid-1950s, has therefore been devoted to finding more efficient ways to generate prime implicants. Two basic approaches, *iterated consensus* and *multiplying* (both given in 1937 by Blake), have emerged from this work. We discuss these approaches in the next two sections.

3.5 Iterated Consensus

Theorem A.2.3 guarantees that any SOP formula is transformed into a syllogistic formula by repeated application of the following rule:

> **If the formula contains a pair r, s of terms whose consensus $c(r, s)$ exists and is not included in any term of the formula, then adjoin $c(r, s)$ to the formula.**

This method, usually called "iterated consensus," was presented in Blake's dissertation. It was re-discovered by Samson and Mills [174], by Quine [163] and by Bing [8, 9]. To apply it, we begin with an SOP formula F. At each application of the rule cited above, we determine, for a pair r, s of terms in F, whether $c(r, s)$ exists, *i.e.*, whether r and s are opposed in exactly one variable. If so, and if $c(r, s)$ is not included in any term in F, we modify F by adjoining to it the term $c(r, s)$. We persevere in this process until every term-pair (involving adjoined terms as well as those originally in F) has been considered. F is then syllogistic.

The process of iterated consensus terminates in a finite number of steps, because

- if a pair r, s of terms meets the specified condition at a given step, it cannot meet that condition at any subsequent step, and

- the number of pairs of terms to be considered is finite, inasmuch as no more than 3^n terms can be produced from n letters. (*NB:* Some writers put this number at $3^n - 1$, excluding 1 as a term.)

We say that a consensus is *applicable* to the formula from which it is derived if the consensus is not included in any term of that formula.

Example 3.5.1 Let us find $BCF(f)$ for the function f expressed by the formula

$$F = w'x'yz + xy'z + wy'z' + xyz' + wx'z' . \qquad (3.5)$$

An organized way to consider all pairs of terms is to compare each term with all those that precede it, adjoining applicable consensus-terms to the end of the formula. To help with bookkeeping, each term should be marked after it has been compared with all preceding terms. The process ends when the consensus of the last term with each of its predecessors either does not exist or is not applicable. The first few stages in the evolution of F, using the foregoing scheme, are shown below:

$$
\begin{aligned}
F &= w'x'yz + xy'z + wy'z' + xyz' + wx'z' \\
F &= w'x'yz + xy'z + wy'z' + xyz' + wx'z' + wxy' \\
F &= w'x'yz + xy'z + wy'z' + xyz' + wx'z' + wxy' + wxz' .
\end{aligned}
$$

The final formula is

$$ F = w'x'yz + xy'z + wy'z' + xyz' + wx'z' + wxy' + wxz' + wyz' + wz' . $$

F is now in syllogistic form; hence,

$$ BCF(f) = ABS(F) = w'x'yz + xy'z + xyz' + wxy' + wz'. \qquad (3.6) $$

It follows from Lemma A.2.1 that the set of formulas formally included in an SOP formula F is not changed by removal from F of absorbed terms. Thus the work of iterated consensus may be simplified at any stage by deletion of absorbed terms as they are noticed. It is not necessary in hand-calculation to be systematic; absorptions missed while generating consensus-terms may be carried out in a final absorption-step.

A variety of iterated-consensus procedures have been investigated, based on specializations of

- the initial SOP formula or

- the interleaving of consensus-generation and absorption.

We now describe two of the more important of such procedures, *viz.*, Quine's method and successive extraction.

3.5.1   Quine's method

A procedure given by Quine [161] is a specialization of iterated consensus in which stages of consensus-generation alternate in a fixed way with stages of absorption. Let us consider again the function given in Example 3.5.1. Quine's method is conveniently explained by organizing the work as shown in Table 3.1. The minterms of f, each containing n literals, are written in column 1. The consensus-terms derived from column 1 (containing $n - 1$ literals each) are written in column 2, after which the terms in column 1 absorbed by terms in column 2 are checked, indicating deletion. This process is carried out repeatedly, column 3 being derived from column 2, column 4 from column 3, and so on. Each column generates the succeeding column by consensus and may suffer absorption of some of its terms by the succeeding column. The terms surviving unchecked (*i.e.*, unabsorbed) are the terms of BCF(f).

$wx'y'z'$ ✓		
$w'x'yz$	$wx'z'$ ✓	
$w'xy'z$ ✓	$wy'z'$ ✓	
$w'xyz'$ ✓		
$wx'yz'$ ✓		
$wxy'z'$ ✓		
$wxy'z$ ✓	$xy'z$	wz'
$wxyz'$ ✓	wxy'	
	xyz'	
	wyz' ✓	
	wxz' ✓	

Table 3.1: Organization of work for Quine's method.

The operations in each pass are on terms of fixed length and all operations (consensus and absorption) derive from the single rule $x'p + xp = p$, where x is a single letter and p is a term. Thus the work involves simple operations organized in convenient stages. Quine's method is made simpler by grouping the minterms in column 1, as shown in Table 3.1, according to the number of their complemented literals. The consensus-operation is then possible only between terms in vertically adjacent groups; this property is inherited by all subsequent columns, as shown in Table 3.1. Further simplification results from introducing binary [134] or octal [6] notation.

The computational advantages of Quine's method are offset by the necessity to begin with a listing of the minterms of f, and by the need therefore to process a large number of terms.

3.5.2 Successive extraction

Blake [10] observed that iterated consensus can be carried out *letter by letter*. This method is called "successive extraction" in the Reusch & Detering survey [167] and is commonly credited to Tison [204, 205]. Suppose an SOP formula involves the letters $a, b, c, \ldots$. To generate an equivalent syllogistic formula by successive extraction, one first adjoins to the formula all applicable consensuses arising from pairs of terms opposed in the letter a, *i.e.*, from pairs of the form $a'p$ and aq, where p and q are terms not involving a. Using the resulting formula as a basis, one then adjoins all applicable consensuses arising from terms opposed in the letter b. This process is repeated until all letters are exhausted.

The method of successive extraction typically generates fewer consensus-terms than does the general method of iterated consensus; however successive extraction (unlike the general method) may require that a given pair of terms be compared more than once.

3.6 Multiplication

Step 1 of Blake's procedure for generating $BCF(f)$ is to find a syllogistic formula for f. The following procedure is guaranteed by Theorem A.2.1 to produce such a formula:

> **Express f as a conjunction of syllogistic formulas and then multiply out to obtain an SOP formula, using the distributive laws and dropping duplicate literals.**

An alterm (disjunction of literals) is clearly syllogistic; hence a syllogistic formula may be produced by multiplying out a conjunction of alterms, *i.e.*, a POS formula. Blake states that the latter method was known to C.S. Peirce [152] and his students. That technique is now frequently attributed to Nelson [147]; Blake's more general technique of multiplying out a conjunction of syllogistic formulas was re-discovered (specialized to a conjunction of Blake canonical forms) by Samson and Mills [174] and by House and Rado [88].

Example 3.6.1 Let us find the prime implicants of the Boolean function f expressed by

$$f = a'd + abc' + ac'd' . \tag{3.7}$$

To apply the method of multiplying, we must first convert formula (3.7) to a conjunction of syllogistic formulas. Let us adopt the specialized tactic of converting (3.7) to a POS formula:

$$f = [a' + bc' + c'd'][a + d] \tag{3.8}$$
$$f = [a' + c'][a' + c + b + d'][a + d] . \tag{3.9}$$

The foregoing conversion is carried out by repeated application of Corollary 2.8.1 (the dual form of Boole's expansion theorem). Specifically, formula (3.8) results from expanding (3.7) with respect to the argument a, *i.e.*,

$$f(a, b, c, d) = [a' + f(1, b, c, d)][a + f(0, b, c, d)] , \tag{3.10}$$

while formula (3.9) is derived by expanding the first factor of (3.8) with respect to the argument c. Multiplying out (3.9) produces the syllogistic formula

$$F = a'd + a'cd + a'bd + a'c'd + abc' + bc'd + ac'd' . \tag{3.11}$$

The terms in (3.11) surviving absorption are the prime implicants of f; thus

$$BCF(f) = ABS(F) = a'd + abc' + bc'd + ac'd' . \tag{3.12}$$

The same result can be obtained by multiplying out (3.8) rather than (3.9), because (as we now show) each of the two factors of (3.8) is syllogistic. By Theorem A.2.3, an SOP formula that is not syllogistic must contain a pair p, q of terms, having exactly one opposition, such that the consensus $c(p, q)$ is not formally included in the formula. Neither of the factors of (3.8) contains such a pair of terms; hence, each factor is syllogistic.

3.6.1 Recursive multiplication

A Blake canonical form is syllogistic. Hence, it follows from Theorem A.2.1 that multiplying out a conjunction of Blake canonical forms produces a syllogistic formula. If absorption is then carried out, the result is a Blake canonical form. Thus $BCF(f)$ may be generated recursively.

Theorem 3.6.1 *Let f be a Boolean function and let x be one of its arguments. Then the Blake canonical form of f is given by*

$$BCF(f) = ABS((x' + BCF(f/x)) \times (x + BCF(f/x'))) , \tag{3.13}$$

where $\times$ (cf. Section A.2) denotes the term-by-term product of SOP formulas.

Proof.   The set $\{x', x\}$ is orthonormal; hence, by Theorem 2.15.1, $f = (x' + f/x) \cdot (x + f/x')$. Thus,

$$BCF(f) = ABS(BCF(x' + f/x) \times BCF(x + f/x')) . \qquad (3.14)$$

The non-vacuous arguments of f/x are disjoint from x; therefore

$$BCF(x' + f/x) = BCF(x') + BCF(f/x) = x' + BCF(f/x) ;$$

similarly,

$$BCF(x + f/x') = x + BCF(f/x') .$$

Equation (3.13) thus follows from (3.14). $\square$

Assuming that F is an SOP formula, $BCF(F)$ is produced by the following recursive procedure:

Rule 1. If the term 1 appears in F, then $BCF(F) = 1$.

Rule 2. If $F = 0$ or F has a single term, then $BCF(F) = F$.

Rule 3. Otherwise,

$$BCF(F) = ABS(\ (x' + BCF(F/x)) \times (x + BCF(F/x')) \ )$$

where x is an argument explicit in F and F/x is expressed by a formula not involving x.

The efficiency of the foregoing procedure may be improved by restricting the scope of the operator ABS. Applying the distributive laws, equation (3.13) may be expressed equivalently as

$$BCF(F) = ABS(G + H) \qquad (3.15)$$

where

$$
\begin{aligned}
G &= (x' \times BCF(F/x')) + (x \times BCF(F/x)) & (3.16) \\
H &= BCF(F/x) \times BCF(F/x') . & (3.17)
\end{aligned}
$$

It is not possible for any term in G to absorb any other term in G; nor can any term in G absorb a term in H. Thus the following absorptions suffice in (3.15):

- absorptions within H; and
- absorptions of G-terms by H-terms.

Let us define a relative-absorption operator, $ABSREL$, on two SOP formulas P and Q, as follows:

$$ABSREL(P,Q) \quad = \quad \text{the formula constructed from P} \\ \text{by removing all terms} \\ \text{absorbed by terms of Q.}$$

Then $BCF(F)$ is expressed recursively by

$$BCF(F) = ABS(H) + ABSREL(G, ABS(H)), \tag{3.18}$$

where G and H are defined by (3.16) and (3.17).

A variation of the foregoing development replaces term-by-term multiplication by Boolean multiplication of arbitrary form. Define SOP formula I to be the Blake canonical form of H as defined above. Then $I = BCF(BCF(F/x) \times BCF(F/x'))$, whence I may be expressed by

$$I = BCF((F/x) * (F/x')), \tag{3.19}$$

where $*$ refers to the product of Boolean functions, the form being irrelevant. Equation (3.19) then takes the simplified form

$$BCF(F) = I + ABSREL(G, I). \tag{3.20}$$

3.6.2 Combining multiplication and iterated consensus

We consider in this section a variant of recursive multiplication that is useful for hand-computation. It is based on the observation that $BCF(F)$ may be computed rapidly by hand if the number of terms in F is relatively small. Beginning with an SOP formula F,

Rule 1. If F is a relatively simple formula, calculate $BCF(F)$ using iterated consensus.

Rule 2. Otherwise,

$$BCF(F) = ABS(\ (x' + BCF(F/x)) \times (x + BCF(F/x'))\ ),$$

where x is an argument appearing with relatively high frequency in F and F/x is expressed by a formula not involving x.

The foregoing procedure is a guide to calculation rather than an algorithm, because Rule 1 requires a decision based on simplicity, a property difficult to quantify. For hand-calculation, however, we have found this procedure to be markedly faster and less conducive to error than any other method, especially when applied to functions yielding large numbers of prime implicants. An analysis of the efficiency of this method in comparison with other methods is given in [25].

Example 3.6.2 Let us apply the foregoing procedure to calculate the Blake canonical form of the formula

$$
\begin{aligned}
F \;=\; & a'c'd' + abd'e' + b'ce + b'c'd'e + abc'd + \\
& + b'c'de' + a'bcd + acd'e' + bc'de \;.
\end{aligned}
\tag{3.21}
$$

We decide that it is not convenient to calculate $BCF(F)$ using Rule 1, and we note that no argument appears in more terms than does c; hence we calculate $BCF(F)$ using Rule 2:

$$
BCF(F) = ABS(\; (c' + BCF(F/c)) \times (c + BCF(F/c')) \;)
\tag{3.22}
$$

where

$$
\begin{aligned}
F/c \;&=\; abd'e' + b'e + a'bd + ad'e' \\
F/c' \;&=\; a'd' + abd'e' + b'd'e + abd + b'de' + bde \;.
\end{aligned}
$$

We decide that the formula F/c is simple enough for application of Rule 1, yielding

$$
BCF(F/c) = b'e + a'bd + ad'e' + a'de + ab'd' \;.
$$

We now decide that it is not convenient to calculate $BCF(F/c')$ using Rule 1, and (noting that the variable d appears with maximal frequency in F/c') we apply Rule 2, *i.e.*,

$$
BCF(F/c') = ABS(\; (d' + BCF((F/c')d)) \times (d + BCF((F/c')d')) \;),
$$

where

$$
\begin{aligned}
(F/c')/d \;&=\; F/c'd \;=\; ab + b'e' + be \\
(F/c')/d' \;&=\; F/c'd' \;=\; a' + abe' + b'e \;.
\end{aligned}
$$

These formulas are relatively simple, hence, we calculate their Blake canonical forms using Rule 1, with the following results:

$$
\begin{aligned}
BCF(F/c'd) \;&=\; ab + b'e' + be + ae' \\
BCF(F/c'd') \;&=\; a' + b'e + be'
\end{aligned}
$$

Thus $BCF(F/c')$ is given by the formula

$$ABS((d' + ab + b'e' + be + ae') \times (d + a' + b'e + be'))$$

which yields

$$BCF(F/c') = a'd' + b'd'e + bd'e' + abd + abe' + \\ + b'de' + a'b'e' + bde + a'be + ade' .$$

We now continue the computation (3.22):

$$BCF(F) = ABS((c' + b'e + a'bd + ad'e' + a'de + ab'd') \times \\ (c + a'd' + b'd'e + bd'e' + abd + abe' + \\ + b'de' + a'b'e' + bde + a'be + ade')) .$$

The result,

$$BCF(f) = b'd'e + abc'd + b'c'de' + bc'de + ac'de' + a'c'd' + \\ + a'b'c'e' + a'bc'e + bc'd'e' + abc'e' + b'ce + \\ + a'bcd + a'bde + acd'e' + abd'e' + a'cde + ab'cd' ,$$

requires 31 "intelligent" multiplications and 3 deletions. By intelligent multi-plications, we mean those that make use of the identities $(p+q)(p+r) = p+qr$ and $(ps + q)(p+r) = ps + pq + qr$ to minimize subsequent absorptions. The more commonly-used multiplying technique, on the other hand, begins with a transformation of (3.21) to POS form, a simple example of which is

$$f = (a + b + c' + e)(b + c + d' + e')(a' + b + c + d + e) \\ (a' + c' + d' + e)(a' + b' + d + e')(a + b' + c + d' + e) \\ (a + b' + c' + d)(a' + b' + c' + e') .$$

Performing the $\times$-operation and absorption on this form (once again assum-ing intelligent multiplications) requires 117 multiplications and 44 deletions.

3.6.3 Unwanted syllogistic formulas

The fact that multiplying out a POS formula produces a syllogistic result may sometimes be disadvantageous. Suppose that our object is to obtain an SOP representation (not necessarily syllogistic) of the complement of f. If we apply De Morgan's laws to an SOP formula for f, we obtain a POS formula for f', which we may then multiply out to produce an SOP formula for f'. Multiplying out a POS formula for f', however, produces a syllogistic result, which may be a more complex formula than one wishes. Let us recall Example 2.14.2. The complement of the formula $f = vx'y + wxz'$, obtained by De Morgan's laws, was shown in that example to be

$$f' = v'w' + v'x' + v'z + w'x + xz + w'y' + x'y' + y'z \, .$$

This formula is in fact $BCF(f')$, inasmuch as it is the result of multiplying out a POS formula and there are no terms in the formula that are absorbed by other terms. The simpler formula $f' = v'x' + x'y' + w'x + xz$ is obtained by expansion-techniques discussed in Section 2.14.

Exercises

1. Express in Blake canonical form:

$$bc'de + ab'c'd + acde + a'b'ce + ab'cd' + b'cde' + \\ + a'bde' + ac'de' + bcd'e' + abce'$$

2. Express in Blake canonical form:

$$AE'F' + DEF + BDE' + A'B'E'F' + BCD'F + \\ + BDEF' + B'D'EF + BD'E'F'$$

3. Express in Blake canonical form:

$$A'BD + AB'D'E + BCD'E + ABDE' + A'E'F + \\ + ADE + A'BD'E' + A'B'E'F'$$

Chapter 4

Boolean Analysis

In this chapter we consider methods for analyzing systems of Boolean equations. These methods are of central importance in Boolean reasoning. We first consider ways in which systems of Boolean equations may be related. In particular, we define consequents, antecedents, equivalents, and solutions of Boolean systems. We also discuss several processes useful for Boolean reasoning. Among such processes are the reduction of a system of equations to a single equivalent equation, the elimination of a variable from an equation, the detection of redundant variables in an interval, and the substitution of an expression for a variable in a Boolean formula.

A problem that arises in diverse applications is to decide whether a given SOP formula is equivalent to the 1-formula. We consider how such a decision may be made, and apply the results to the problem of finding a near-minimal SOP formula for a given Boolean function.

4.1 Review of Elementary Properties

We repeat for convenient reference some equivalences developed in Chapter 2. For elements a, b, c in a Boolean algebra,

$$a \leq b \quad \Longleftrightarrow \quad ab' = 0 \tag{4.1}$$

$$a \leq b \leq c \quad \Longleftrightarrow \quad ab' + bc' = 0 \tag{4.2}$$

$$a = b \quad \Longleftrightarrow \quad a \oplus b = 0 \tag{4.3}$$

$$a = 0 \text{ and } b = 0 \quad \Longleftrightarrow \quad a + b = 0 \tag{4.4}$$

$$a = 1 \text{ and } b = 1 \quad \Longleftrightarrow \quad ab = 1 \tag{4.5}$$

Equivalences (4.4) and (4.5) extend readily to more than three variables; thus

$$a = 0 \text{ and } b = 0 \text{ and } c = 0 \quad \Longleftrightarrow \quad a + b + c = 0$$

is an obvious generalization of (4.4).

Boole's expansion theorem (Theorem 2.8.1), together with property (4.2), establishes

Proposition 4.1.1 (Schröder [178]) *The statements*

$$f(x, y, \ldots) = 0$$

and

$$f(0, y, \ldots) \leq x \leq f'(1, y, \ldots)$$

are equivalent.

4.2   Boolean Systems

An *n-variable Boolean system* on a Boolean algebra **B** is a collection

$$
\begin{aligned}
p_1(X) &= q_1(X) \\
&\vdots \\
p_k(X) &= q_k(X)
\end{aligned}
\tag{4.6}
$$

of simultaneously-asserted equations. The p_i and q_i are n-variable Boolean functions on **B**; X denotes the vector $(x_1, x_2, \cdots, x_n)$. We have defined a Boolean system to consist entirely of equations; inclusions are readily transformed into equations, however, via the equivalence (4.1).

Given any substitution $A \in \mathbf{B}^n$ for X, a truth-value is assigned to a Boolean system $S(X)$ as follows: $S(A)$ is true in case each of its component equations is an identity; otherwise $S(A)$ is false. A Boolean system, and thus a Boolean equation, is therefore a *predicate* (*cf.* the discussion in Section 1.2).

4.2.1 Antecedent, Consequent, and Equivalent Systems

Let $S_1(X)$ and $S_2(X)$ be two n-variable Boolean systems on **B**. We say that $S_1(X)$ is an *antecedent* of $S_2(X)$, written $S_1(X) \implies S_2(X)$, in case every substitution for X that causes $S_1(X)$ to be true also causes $S_2(X)$ to be true; we say in this case also that $S_2(X)$ is a *consequent* of $S_1(X)$. Two Boolean systems $S_1(X)$ and $S_2(X)$ are said to be *equivalent*, written $S_1(X) \iff S_2(X)$, if each is a consequent of the other.

4.2.2 Solutions

A system having the specialized form

$$
\begin{aligned}
x_1 &= b_1 \\
&\vdots \\
x_n &= b_n \, ,
\end{aligned}
\tag{4.7}
$$

where $b_1, b_2, \cdots, b_n$ are elements of **B**, is called a *solution* of a system $S(X)$ provided (4.7) implies (*i.e.,* is an antecedent of) $S(X)$. Thus (4.7) is a solution of $S(X)$ in case the substitutions defined by (4.7) cause each of the equations in $S(X)$ to become an identity; it is sometimes convenient to refer to the vector $(b_1, b_2, \cdots, b_n)$ itself as a solution. A Boolean system is said to be *consistent* if it has at least one solution; otherwise, it is said to be *inconsistent*. Methods for constructing solutions are discussed in Chapter 7.

4.3 Reduction

Unlike a system of equations in "ordinary" algebra, a Boolean system may be reduced to a single equivalent equation.

Theorem 4.3.1 (Boole [13], Chapter VIII) *The Boolean system* (4.6) *is equivalent to the single equation*

$$
f(X) = 0,
\tag{4.8}
$$

where f is defined by

$$
f = \sum_{i=1}^{k} (p_i \oplus q_i) \, .
\tag{4.9}
$$

Proof. System 4.8 is equivalent, by property (4.3), to the system

$$p_1(X) \oplus q_1(X) = 0$$
$$\vdots$$
$$p_k(X) \oplus q_k(X) = 0$$

which is in turn equivalent, by property (4.4), to the single equation (4.8), where f is the Boolean function defined by (4.9). $\square$

By similar reasoning, invoking (4.5) instead of (4.4), we arrive at

Corollary 4.3.1 *The system* (4.6) *is equivalent to the single equation*

$$F(X) = 1, \tag{4.10}$$

where F is a Boolean function defined by

$$F = \prod_{i=1}^{k} (p_i \oplus q_i)'. \tag{4.11}$$

Any Boolean system can therefore be reduced to a single equivalent equation whose right-hand side is either zero or one. More generally, as shown in Theorem 4.5.1 (Poretsky's Law of Forms), the right-hand side may be any preassigned Boolean function.

Example 4.3.1 The system

$$ax = b + y$$
$$ab \leq ax' + y'$$

is equivalent to the system

$$ab'xy' + a'b + a'y + bx' + x'y = 0$$
$$ab(a'y + xy) = 0$$

which is equivalent, in turn, to the single equation

$$ab'xy' + a'b + a'y + bx' + x'y + abxy = 0.$$

$\square$

Example 4.3.2 Suppose an AND-gate to have inputs x_1 and x_2 and output z_1. The behavior of the gate is described by any of the three equivalent statements below.

$$x_1 x_2 = z_1 \tag{4.12}$$
$$x_1' z_1 + x_2' z_1 + x_1 x_2 z_1' = 0 \tag{4.13}$$
$$x_1' z_1' + x_2' z_1' + x_1 x_2 z_1 = 1 \tag{4.14}$$

□

4.4 The Extended Verification Theorem

We discuss in this section a result, due to Löwenheim [124] and Müller [142], which enables an implication between two Boolean equations to be transformed into an equivalent Boolean inclusion. The presentation in this section is adapted from that of Rudeanu [172].

Let s be a single element of **B** and let $V = (v_1, v_2, \cdots, v_n)$ be a vector on **B**, *i.e.*, $s \in$ **B** and $V \in$ **B**^{n}. Then sV and Vs are defined by

$$sV = Vs = (sv_1, sv_2, \cdots, sv_n).$$

Lemma 4.4.1 *Let f:* **B**^{n}*$\longrightarrow$**B** *be a Boolean function and let A be an element of* **B**^{n} *such that $f(A) = 0$. Then*

$$f(Af(X) + Xf'(X)) = 0 \qquad (\forall X \in \mathbf{B}^n). \tag{4.15}$$

Proof. By Boole's expansion theorem (Theorem 2.8.1),

$$f(G(u)) = u' \cdot f(G(0)) + u \cdot f(G(1));$$

thus, setting $G(u) = Au + Xu'$ and $u = f(X)$,

$$f(Af(X) + Xf'(X)) = f'(X)f(X) + f(X)f(A).$$

Each term on the right-hand side of the foregoing equation has the value zero, for any $X \in \mathbf{B}^n$, verifying (4.15). □

Theorem 4.4.1 (Extended Verification Theorem) *Let f :* **B**$^n \longrightarrow$ **B** *and g:* **B**$^n \longrightarrow$ **B** *be Boolean functions, and assume equation $f(X) = 0$ to be consistent. Then the following statements are equivalent:*

(a) $\quad f(X) = 0 \implies g(X) = 0 \quad (\forall X \in \mathbf{B}^n)$

(b) $\quad\quad\quad g(X) \leq f(X) \quad\quad\quad (\forall X \in \mathbf{B}^n)$

(c) $\quad\quad\quad g(X) \leq f(X) \quad\quad\quad (\forall X \in \{0,1\}^n).$

Proof.

(a) $\Longrightarrow$ (b): Let $A \in \mathbf{B}^n$ be a solution of $f(X) = 0$, *i.e.*, let $f(A) = 0$ be an identity. Then $g(A) = 0$ because of the assumed implication. By Lemma 4.4.1, the equation $f(X f'(X) + A f(X)) = 0$ is satisfied for any $X \in \mathbf{B}^n$; hence $g(X f'(X) + A f(X)) = 0$ is also satisfied. Thus

$$f'(X)g(X) + f(X)g(A) = f'(X)g(X) = 0 ,$$

i.e., $g(X) \leq f(X)$, proving (b).

(b) $\Longrightarrow$ (c): Immediate.

(c) $\Longrightarrow$ (a): The functions f and g are Boolean; hence they may be written in minterm canonical form, *i.e.*,

$$f(X) = \sum_{K \in \{0,1\}^n} f(K) X^K \quad \text{and} \quad g(X) = \sum_{K \in \{0,1\}^n} g(K) X^K$$

for all $X \in \mathbf{B}^n$. Assume (c), *i.e.*, assume $g(K) \leq f(K)$ for all $K \in \{0,1\}^n$, and let $A \in \mathbf{B}^n$ be a solution of $f(X) = 0$. Then

$$f(X) = \sum_{K \in \{0,1\}^n} f(K) A^K = 0 ,$$

which implies that $f(K) A^K = 0$ for all $K \in \{0,1\}^n$, and therefore that $g(K) A^K = 0$ for all $K \in \{0,1\}^n$. Thus $g(A) = 0$, proving (a). $\Box$

Corollary 4.4.1 *Let $f: \mathbf{B}^n \longrightarrow \mathbf{B}$ and $g: \mathbf{B}^n \longrightarrow \mathbf{B}$ be Boolean functions, and assume the equation $f(X) = 0$ to be consistent. Then the following statements are equivalent:*

(a) $\quad f(X) = 0 \Longleftrightarrow g(X) = 0 \quad (\forall X \in \mathbf{B}^n)$

(b) $\quad\quad\quad g(X) = f(X) \quad\quad (\forall X \in \mathbf{B}^n)$

(c) $\quad\quad\quad g(X) = f(X) \quad\quad (\forall X \in \{0,1\}^n) .$

Proof. Immediate from Theorem 4.4.1 and the definition of equivalent systems. $\Box$

4.5 Poretsky's Law of Forms

Theorem 4.3.1 and its corollary enable us to reduce a Boolean system to one of the equivalent forms $f(X) = 0$ or $F(X) = 1$. Suppose, however, that we wish the right-hand side to be something other than 0 or 1. Poretsky [159] showed that once a Boolean system has been reduced to the form $f(X) = 0$, it may be re-expressed equivalently in the form $g(X) = h(X)$, where h is any specified Boolean function. The function g associated with a given h is determined uniquely by the following theorem.

Theorem 4.5.1 (Poretsky's Law of Forms) *Let $f, g, h : \mathbf{B}^n \longrightarrow \mathbf{B}$ be Boolean functions and suppose the equation $f(X) = 0$ to be consistent. Then the equivalence*

$$f(X) = 0 \qquad \Longleftrightarrow \qquad g(X) = h(X) \tag{4.16}$$

holds for all $X \in \mathbf{B}^n$ if and only if

$$g = f \oplus h . \tag{4.17}$$

Proof. Equivalence (4.16) may be written in the form

$$f(X) = 0 \qquad \Longleftrightarrow \qquad g(X) \oplus h(X) = 0 ,$$

which is equivalent by Corollary 4.4.1 to (4.17). $\square$

Example 4.5.1 Suppose a Boolean function, g, is sought having the property that the equation $x_1 x_2' + x_3 = 0$ is equivalent to $g = x_2 x_3$. The first equation is consistent (a solution, for example, is $(x_1, x_2, x_3) = (0, 0, 0)$); hence, g is determined uniquely by (4.17), i.e.,

$$\begin{aligned} g &= (x_1 x_2' + x_3) \oplus (x_2 x_3) \\ &= x_2'(x_1 + x_3) . \end{aligned}$$

$\square$

4.6 Boolean Constraints

Given a Boolean algebra $\mathbf{B}$, a *constraint* on a vector $X = (x_1, x_2, \cdots, x_n)$ is a statement confining X to lie within a subset of $\mathbf{B}^n$. A constraint is therefore a predicate that is true provided X is a member of the subset. The operation of the AND-gate of Example 4.3.2, for instance is specified by the constraint

$$(x_1, x_2, z_1) \in \{(0,0,0), (0,1,0), (1,0,0), (1,1,1)\} , \tag{4.18}$$

where $\mathbf{B} = \{0, 1\}$ and $X = (x_1, x_2, z_1)$.

Two constraint-statements on the vector X are *equivalent* if they are equivalent as predicates, *i.e.*, if they confine X to the same subset of $\mathbf{B}^n$. Thus statement (4.18) is equivalent to equation (4.12), as well as to equations (4.13) and (4.14).

An *identity* on $X = (x_1, x_2, \cdots, x_n)$ is a constraint equivalent to the statement

$$X \in \mathbf{B}^n .$$

An identity, in other words, is a constraint that doesn't really do any constraining. The constraints $x + y = x + x'y$ and $x'y \leq y$, for example, are both identities on (x, y).

A constraint on $X = (x_1, x_2, \cdots, x_n)$ will be called a *Boolean constraint* if it is equivalent to a Boolean equation, *i.e.*, if it can be expressed by the equation

$$f(X) = 0 , \tag{4.19}$$

where $f: \mathbf{B}^n \longrightarrow \mathbf{B}$ is a Boolean function. If $\mathbf{B} = \{0, 1\}$, then every constraint on X is Boolean; if $\mathbf{B}$ is larger than $\{0, 1\}$, however, then not all constraints on X are Boolean.

Example 4.6.1 Suppose that $\mathbf{B} = \{0, 1, a', a\}$. Then the constraint

$$(x, y) \in \{(0, 0), (a, 0)\} \tag{4.20}$$

is Boolean because it is equivalent to the Boolean equation

$$a'x + y = 0 , \tag{4.21}$$

i.e., the set of solutions of (4.21) is $\{(0, 0), (a, 0)\}$ for $\mathbf{B} = \{0, 1, a', a\}$. $\square$

Example 4.6.2 Suppose again that $\mathbf{B} = \{0, 1, a', a\}$. The constraint

$$(x, y) \in \{(0, 0), (a, 1)\} , \tag{4.22}$$

is not a Boolean constraint, *i.e.*, it is not equivalent to a Boolean equation. To show this, let us assume that there is a two-variable Boolean function f whose solution-set is $\{(0, 0), (a, 1)\}$. Then

$$x = 0 \text{ and } y = 0 \implies f = 0$$
$$x = a \text{ and } y = 1 \implies f = 0 ,$$

i.e.,

$$x + y = 0 \implies f = 0$$
$$ax' + a'x + y' = 0 \implies f = 0 .$$

By Theorem 4.4.1, the extended verification theorem, the latter implications are equivalent to the inclusions

$$f \leq x + y$$
$$f \leq ax' + a'x + y' .$$

These inclusions are equivalent together to the single inclusion

$$f \leq g , \tag{4.23}$$

where the Boolean function g is defined by

$$\begin{aligned} g &= (x + y) \cdot (ax' + a'x + y') \\ &= a'x + xy' + ax'y . \end{aligned}$$

Invoking the extended verification theorem again shows that the inclusion (4.23) is equivalent to the implication

$$g(x, y) = 0 \implies f(x, y) = 0 \qquad (\forall (x, y) \in \mathbf{B}^2) ;$$

thus every solution of $g(x, y) = 0$ is also a solution of $f(x, y) = 0$. Trying all values for (x, y) in $\mathbf{B}^2$ shows that the solution-set of $g(x, y) = 0$ is $\{(0, 0), (0, a'), (a, a), (a, 1)\}$. Thus the solution-set of $f(x, y) = 0$ must contain these four elements, contradicting the assumption that its solution-set is $\{(0, 0), (a, 1)\}$. Hence $(x, y) \in \{(0, 0), (a, 0)\}$ is not a Boolean constraint. $\square$

4.7 Elimination

A fundamental process in Boolean reasoning is that of *elimination*. To eliminate a variable x from a Boolean equation means to derive another Boolean equation that expresses all that can be deduced from the original equation without reference to x. The central fact concerning elimination was announced by Boole [13, Chapt. VII, Proposition I] as follows:

If $f(x) = 0$ be any logical equation involving the class symbol x, with or without other class symbols, then will the equation

$$f(1)f(0) = 0$$

be true, independently of the interpretation of x; and it will be the complete result of the elimination of x from the above equation.

In other words, the elimination of x from any given equation, $f(x) = 0$, will be effected by successively changing in that equation x into 1, and x into 0, and multiplying the two resulting equations together.

Similarly, the complete result of the elimination of any class symbol, x, y, &c., from any equation of the form $V = 0$, will be obtained by completely expanding the first member of that equation in constituents of the given symbols, and multiplying together all the coefficients of those constituents, and equating the product to 0.

Let $X = (x_1, x_2, \ldots, x_m)$, let $Y = (y_1, y_2, \ldots, y_n)$, and let $f: \mathbf{B}^{m+n} \longrightarrow \mathbf{B}$ be a Boolean function. Following Boole, we define the *resultant of elimination* of X from the equation $f(X, Y) = 0$ to be the equation

$$\prod_{A \in \{0,1\}^m} f(A, Y) = 0 . \tag{4.24}$$

It follows that the resultant of elimination of X from $F(X, Y) = 1$ is the equation

$$\sum_{A \in \{0,1\}^m} F(A, Y) = 1 . \tag{4.25}$$

To demonstrate that the resultant of elimination is Boole's "complete result," we must show that an equation $h(Y) = 0$ is a consequent of $f(X, Y) = 0$ if and only if it is a consequent of the resultant (4.24).

Theorem 4.7.1 *Let* $X = (x_1, x_2, \ldots, x_m)$ *and* $Y = (y_1, y_2, \ldots, y_n)$ *be disjoint argument-vectors, and let* $f: \mathbf{B}^{m+n} \longrightarrow \mathbf{B}$ *be a Boolean function. For any Boolean function* $h: \mathbf{B}^n \longrightarrow \mathbf{B}$, *the implications*

$$f(X, Y) = 0 \quad \Longrightarrow \quad h(Y) = 0 \quad (\forall X \in \mathbf{B}^m) \tag{4.26}$$

$$\prod_{A \in \{0,1\}^m} f(A, Y) = 0 \quad \Longrightarrow \quad h(Y) = 0 \tag{4.27}$$

are equivalent.

Proof. If equation $f(X,Y) = 0$ is inconsistent, then (4.26) and (4.27) are both true because their premises are false. Otherwise a repeated application of the Verification Theorem and its variants yields successively the following equivalent forms of (4.26):

$$
\begin{aligned}
h(Y) &\le & f(X,Y) & & \forall X \in \mathbf{B}^m \\
h(Y) &\le & f(A,Y) & & \forall A \in \{0,1\}^m \\
h(Y) &\le & \textstyle\prod_{A\in\{0,1\}^m} f(A,Y)
\end{aligned}
$$

The latter inclusion is equivalent to (4.27). $\Box$

Example 4.7.1 The AND-gate of Example 4.3.2 is characterized by either of the equations $f(x_1, x_2, z_1) = 0$ or $F(x_1, x_2, z_1) = 1$, the functions f and F being defined by

$$
\begin{aligned}
f &= x_1' z_1 + x_2' z_1 + x_1 x_2 z_1' & (4.28) \\
F &= x_1' z_1' + x_2' z_1' + x_1 x_2 z_1 . & (4.29)
\end{aligned}
$$

Let us eliminate x_2. Applying the definitions (4.24) and (4.25), the resultant of elimination of x_2 is expressed by either of the equations $g(x_1, z_1) = 0$ or $G(x_1, z_1) = 1$, where

$$
\begin{aligned}
g &= f(x_1, 0, z_1) \cdot f(x_1, 1, z_1) & = (x_1' z_1 + z_1)(x_1' z_1 + x_1 z_1') \\
& & = x_1' z_1 \\
G &= F(x_1, 0, z_1) + F(x_1, 1, z_1) & = (x_1' z_1' + z_1') + (x_1' z_1' + x_1 z_1) \\
& & = x_1 + z_1' .
\end{aligned}
$$

All that is known concerning the AND-gate's input x_1 and output z_1, in the absence of knowledge concerning its input x_2, is therefore expressed by any of the following equivalent statements:

$$
\begin{aligned}
x_1' z_1 &= 0 \\
x_1 + z_1' &= 1 \\
z_1 &\le x_1 \\
(x_1, z_1) &\in \{(0,0),(1,0),(1,1)\} .
\end{aligned}
$$

If we eliminate the output-argument z_1 from (4.13), the resultant is

$$
(x_1 x_2)(x_1' + x_2') = 0
$$

i.e.,

$$0 = 0 .$$

The latter constraint allows (x_1, x_2) to be chosen freely on $\{0, 1\}^2$—confirming our expectation that the inputs to a gate are unconstrained if nothing is known concerning the value of the output. $\square$

C.I. Lewis [123, p. 155] has observed that "For purposes of application of the algebra to ordinary reasoning, elimination is a process more important than solution, since most processes of reasoning take place through the elimination of 'middle' terms." Boole [13, p. 99] writes of such terms that it "usually happens in common reasoning, and especially when we have more than one premiss, that some of the elements [in the premiss] are not required to appear in the conclusion. Such elements, or, as they are commonly called, "middle terms," may be considered as introduced into the original propositions only for the sake of that connexion which they assist to establish among the other elements, which are alone designed to enter into the expression of the conclusion."

The following example illustrates the process of reasoning by elimination of such middle terms.

Example 4.7.2 Let us connect the output, z_1, of the AND-gate of Example 4.7.1 to the input of an OR-gate whose second input is labelled x_3 and whose output is labelled z_2. The complete circuit is thus defined by the system

$$\begin{aligned} z_1 &= x_1 x_2 \\ z_2 &= x_3 + z_1 . \end{aligned} \qquad (4.30)$$

The relationship between the circuit's overall output, z_2, and its inputs, x_1, x_2, and x_3 is expressed only implicitly by the foregoing equations. We deduce an explicit relationship by eliminating the "middle term," z_1. To do so, we first reduce the system (4.30) to a single equation, *viz.*, $f(x_1, x_2, x_3, z_1, z_2) = 0$, the Boolean function f being given by

$$f = x_1' z_1 + x_2' z_1 + x_1 x_2 z_1' + x_3 z_2' + z_1 z_2' + x_3' z_1' z_2 .$$

The resultant of elimination of z_1 from $f = 0$ is the equation

$$g(x_1, x_2, x_3, z_2) = 0 , \qquad (4.31)$$

where g is given by (4.24) as follows:

$$
\begin{aligned}
g &= f(x_1, x_2, x_3, 0, z_2) \cdot f(x_1, x_2, x_3, 1, z_2) \\
&= (x_1 x_2 + x_3 z_2' + x_3' z_2) \cdot (x_1' + x_2' + x_3 z_2' + z_2') \\
&= x_3 z_2' + x_1 x_2 z_2' + x_1' x_3' z_2 + x_2' x_3' z_2 \, .
\end{aligned}
$$

Proposition 4.1.1 enables us to express equation (4.31) equivalently in a form which isolates z_2:

$$
g(x_1, x_2, x_3, 0) \leq z_2 \leq g'(x_1, x_2, x_3, 1) \, .
$$

Thus

$$
x_3 + x_1 x_2 \leq z_2 \leq x_3 + x_1 x_2 \, ,
$$

i.e.,

$$
z_2 = x_3 + x_1 x_2 \, .
$$

Suppose now that we wish to express what the system (4.30) tells us about the value of x_3, if we know only the values of x_2 and z_2. To do so, we eliminate x_1 and z_1 from $f(x_1, x_2, x_3, z_1, z_2) = 0$ or, equivalently, we eliminate x_1 from $g(x_1, x_2, x_3, z_2) = 0$. The resultant of the latter elimination is

$$
g(0, x_2, x_3, z_2) \cdot g(1, x_2, x_3, z_2) = 0
$$

i.e.,

$$
x_3 z_2' + x_2' x_3' z_2 = 0 \, ,
$$

which is equivalent to the interval

$$
x_2' z_2 \leq x_3 \leq z_2 \, .
$$

This interval tells us the following about x_3:

1. If $x_2 = 0$ and $z_2 = 1$, then $x_3 = 1$.

2. If $z_2 = 0$, then $x_3 = 0$.

A point of difference between Boolean and other algebras with reference to elimination should be noted. As usual, Boole [13, p. 99] states the matter best: "In the [common] algebraic system we are able to eliminate one symbol from two equations, two symbols from three equations, and generally $n - 1$ symbols from n equations. There thus exists a definite connexion between

the number of independent equations given and the number of symbols of
quantity which it is possible to eliminate from them. But it is otherwise with
the system of Logic. No fixed connexion there prevails between the number
of equations given representing propositions or premises, and the number
of typical symbols of which the elimination can be effected. From a single
equation an indefinite number of such symbols may be eliminated."

4.8   Eliminants

As shown in Section 4.7, the resultant of elimination of variable x_1 from
equation $f(x_1, x_2, \ldots) = 0$ is equation $f(0, x_2, \ldots) \cdot f(1, x_2, \ldots) = 0$; similarly,
the resultant of elimination of x_1 from $f(x_1, x_2, \ldots) = 1$ is $f(0, x_2, \ldots) +
f(1, x_2, \ldots) = 1$.

We call functions $f(0, x_2, \ldots) \cdot f(1, x_2, \ldots)$ and $f(0, x_2, \ldots) + f(1, x_2, \ldots)$,
and their generalizations to more than one eliminated variable, *eliminants*;
they are of central importance in Boolean reasoning. An eliminant (a function) is often needed in situations where the corresponding resultant of elimination (an equation) is not needed; therefore it will prove useful to define
eliminants in a way that is independent of the process of elimination. A computer program performing tasks of Boolean reasoning will spend much of its
time computing eliminants; these functions therefore deserve close study.

Let $f \colon \mathbf{B}^n \longrightarrow \mathbf{B}$ be a Boolean function expressed in terms of arguments
$x_1, x_2, \ldots, x_n$, and let R, S, and T be subsets of $\{x_1, x_2, \ldots, x_n\}$. We define
a Boolean function $ECON(f, T)$ by the following rules:

(i) $ECON(f, \emptyset)$ $= f$
(ii) $ECON(f, \{x_1\})$ $= f(0, x_2, \ldots, x_n) \cdot f(1, x_2, \ldots, x_n)$
(iii) $ECON(f, R \cup S)$ $= ECON(ECON(f, R), S)$

We define another Boolean function, $EDIS(f, T)$, by the rules

(i) $EDIS(f, \emptyset)$ $= f$
(ii) $EDIS(f, \{x_1\})$ $= f(0, x_2, \ldots, x_n) + f(1, x_2, \ldots, x_n)$
(iii) $EDIS(f, R \cup S)$ $= EDIS(EDIS(f, R), S)$

We call $ECON(f, T)$ the *conjunctive eliminant*, and $EDIS(f, T)$ the
disjunctive eliminant, of f with respect to the subset T. We note that if T
is a singleton, *i.e.*, if $T = \{x\}$, then the eliminants of f are related to the

quotients f/x' and f/x (Section 2.15) as follows:

$$ECON(f, \{x\}) = f/x' \cdot f/x \qquad (4.32)$$
$$EDIS(f, \{x\}) = f/x' + f/x \qquad (4.33)$$

Theorem 4.8.1 *Let* $f: \mathbf{B}^n \longrightarrow \mathbf{B}$ *be a Boolean function and let* T *be an* m-*element subset of its argument-set,* $X = \{x_1, x_2, \ldots, x_n\}$. *We assume without loss of generality that* T *comprises the first* m *elements of* X, *i.e., that* $T = \{x_1, \ldots, x_m\}$. *Then* $ECON(f, T)$ *and* $EDIS(f, T)$ *are determined as follows:*

$$ECON(f, T) = \prod_{A \in \{0,1\}^m} f(A, x_{m+1}, \ldots, x_n) \qquad (4.34)$$

$$EDIS(f, T) = \sum_{A \in \{0,1\}^m} f(A, x_{m+1}, \ldots, x_n). \qquad (4.35)$$

Proof. Equation (4.34) is verified for the case $m = 1$ by the definition of the conjunctive eliminant. Suppose (4.34) to hold for $m = k > 1$, and consider the case $m = k + 1$:

$$ECON(f(x_1, \ldots, x_k, x_{k+1}, x_{k+2}, \ldots, x_n), \{x_1, \ldots, x_k, x_{k+1}\})$$
$$= ECON(ECON(f, \{x_1, \ldots, x_k\}), \{x_{k+1}\})$$
$$= ECON(\prod_{A \in \{0,1\}^k} f(A, x_{k+1}, x_{k+2}, \ldots, x_n), \{x_{k+1}\})$$
$$= \prod_{A \in \{0,1\}^k} f(A, 0, x_{k+2}, \ldots, x_n) \cdot \prod_{A \in \{0,1\}^k} f(A, 1, x_{k+2}, \ldots, x_n)$$
$$= \prod_{A \in \{0,1\}^{k+1}} f(A, x_{k+2}, \ldots, x_n)$$

Equation (4.34) thus holds for $m = k + 1$, completing the verification of (4.34). Equation (4.35) is verified by dual computations. $\square$

Example 4.8.1

$$ECON(f(w, x, y, z), \{w, y\}) =$$
$$f(0, x, 0, z) \cdot f(0, x, 1, z) \cdot f(1, x, 0, z) \cdot f(1, x, 1, z)$$
$$EDIS(f(w, x, y, z), \{w, y\}) =$$
$$f(0, x, 0, z) + f(0, x, 1, z) + f(1, x, 0, z) + f(1, x, 1, z).$$

$\square$

Corollary 4.8.1 *Let* $X = (x_1, x_2, \ldots, x_m)$ *and* $Y = (y_1, y_2, \ldots, y_n)$ *be disjoint argument-vectors, and let* $f\colon \mathbf{B}^{m+n} \longrightarrow \mathbf{B}$ *be a Boolean function. Then the resultant of elimination of* X *from* $f(X, Y) = 0$ *is*

$$ECON(f, X) = 0 . \tag{4.36}$$

The resultant of elimination of X *from* $f(X, Y) = 1$ *is*

$$EDIS(f, X) = 1 . \tag{4.37}$$

Calculation of Eliminants. It is clear that either the conjunctive or the disjunctive eliminant of a Boolean function f with respect to a subset T may be expressed by a formula not involving any of the arguments appearing in T. The calculation of such formulas is simplified by application of the results which follow.

Proposition 4.8.1 (Schröder [178], Vol. I, Sect. 21). *If a Boolean function* f *is expressed as*

$$f(x, y, \ldots) = x'p(y, \ldots) + xq(y, \ldots) + r(y, \ldots)$$

then the conjunctive eliminant $ECON(f, \{x\})$ *is given by*

$$ECON(f, \{x\}) = pq + r .$$

Proof. $ECON(f, \{x\}) = f(0, y, \ldots)f(1, y, \ldots) = [p + r][q + r] = pq + r .$

Lemma 4.8.1 *Let* $f\colon \mathbf{B}^n \longrightarrow \mathbf{B}$ *be a Boolean function expressed in terms of arguments* $x, y, \ldots$. *Then*

$$BCF(ECON(f, \{x\})) = \sum (\text{terms of } BCF(f) \text{ not involving } x \text{ or } x') .$$

Proof. The literals x and x' may be factored from the terms of $BCF(f)$ in such a way that f is expressed as

$$f = \sum_{i=1}^{L} x'p_i + \sum_{j=1}^{M} xq_j + \sum_{k=1}^{N} r_j ,$$

where $p_1, \ldots, p_L, q_1, \ldots, q_M, , r_1, \ldots, r_N$ are terms (products) not involving the argument x. Thus $ECON(f, \{x\}) = f(0, y, \ldots)f(1, y, \ldots)$ may be expressed as

$$[\sum_{i=1}^{L} p_i + \sum_{k=1}^{N} r_k] [\sum_{j=1}^{M} q_j + \sum_{k=1}^{N} r_k] = \sum_{i=1}^{L} \sum_{j=1}^{M} p_i q_j + \sum_{k=1}^{N} r_k .$$

Every consensus formed by terms of $BCF(f)$ is absorbed by a term of $BCF(f)$. In particular, every consensus of the form $p_i q_j$ is absorbed by one of the r-terms; thus

$$\sum_{i=1}^{L}\sum_{j=1}^{M} p_i q_j \leq \sum_{k=1}^{N} r_k ,$$

and we conclude that $ECON(f, \{x\}) = \sum_{k=1}^{N} r_k$. Thus $ECON(f, \{x\})$ may be expressed as the portion of $BCF(f)$ that remains after each of its terms that involves x or x' is deleted; let us call this portion G. It remains only to show that G is in Blake canonical form. Suppose not. It is clear that G is absorptive, since it is a fragment of a Blake canonical form; hence it must not be syllogistic. By Theorem A.2.3, therefore, there must be terms s and t in G such that the consensus $c(s, t)$ exists and is not formally included in G. Thus $c(s, t)$, which does not contain x or x', is included in one of the terms dropped from $BCF(f)$ in the formation of G. But each such term contains either x or x', which is a contradiction. $\square$

Theorem 4.8.2 *Let* $f: B^n \longrightarrow B$ *be a Boolean function expressed in terms of arguments* $x, y, \ldots$ *and let* T *be a subset of* $\{x, y, \ldots\}$. *Then*

$$BCF(ECON(f,T)) = \sum \text{terms of } BCF(f) \text{ not involving arguments in } T). \tag{4.38}$$

Proof. By Lemma 4.8.1, (4.38) is valid if $\#T = 1$, *i.e.*, if T is a singleton-set. Suppose (4.38) to be valid if $\#T = k$, and consider the case $\#T = k+1$, *i.e.*, let $T = R \cup \{x\}$, where $\#R = k$ and $x \notin R$. Then

$$
\begin{aligned}
BCF(ECON(f,T)) &= BCF(ECON(ECON(f,R),\{x\})) \\
&= \sum (\text{terms of } BCF(ECON(f,R)) \\
&\quad \text{not involving } x \text{ or } x') \\
&= \sum (\text{terms of} \\
&\quad \left[\begin{array}{l} \sum (\text{terms of } BCF(f) \text{ not} \\ \quad \text{involving arguments in } R) \end{array} \right] \\
&\quad \text{not involving } x \text{ or } x')
\end{aligned}
$$

Thus (4.38) is valid for $T = R \cup \{x\}$. $\square$

Conjunctive eliminants of a Boolean function are therefore found by simple term-deletions, provided the function is expressed in Blake canonical form. The resulting eliminants inherit the property of being in Blake canonical form.

Example 4.8.2 The Boolean function

$$f = wx'y + v'w'x' + xz + xy' + vx'y'$$

is expressed as follows in Blake canonical form:

$$\begin{aligned} BCF(f) &= vy' + w'y' + xy' + xz + wx'y + wyz + v'yz + \\ &\quad + v'x'y + v'w'x' + v'w'z + vwz + vwx' \,. \end{aligned}$$

The conjunctive eliminants expressed below may therefore be constructed by inspection of $BCF(f)$, using Theorem 4.8.2.

$$\begin{aligned} BCF(ECON(f, \{v\})) &= w'y' + xy' + xz + wx'y + wyz \\ BCF(ECON(f, \{v, z\})) &= w'y' + xy' + wx'y \\ BCF(ECON(f, \{x, y\})) &= v'w'z + vwz \\ BCF(ECON(f, \{x, y, z\})) &= 0 \,. \end{aligned}$$

□

Theorem 4.8.3 Let $f: B^n \longrightarrow B$ be a Boolean function expressed in terms of the arguments $x, y, \dots$. Then $EDIS(f, \{x\})$ is obtained from any sum-of-products (SOP) formula for f by replacing x and x', wherever they appear in the formula, by 1.

Proof. As in Proposition 4.8.1, the terms in an SOP formula for f may be segregated into those containing x', those containing x, and those containing neither x' nor x. The literals x' and x may then be factored from the terms in which they appear, to produce an expression of the form

$$f(x, y, \dots) = x'P(y, \dots) + xQ(y, \dots) + R(y, \dots) \,,$$

where P, Q, and R are SOP formulas (some possibly null) not involving x. Hence

$$f(0, y, \dots) = P(y, \dots) + R(y, \dots)$$

and

$$f(1, y, \dots) = Q(y, \dots) + R(y, \dots) \,.$$

By definition,

$$EDIS(f(x, y, \ldots), \{x\}) = f(0, y, \ldots) + f(1, y, \ldots);$$

hence

$$EDIS(f(x, y, \ldots), \{x\}) = P(y, \ldots) + Q(y, \ldots) + R(y, \ldots).$$

Thus $EDIS(f(x, y, \ldots), \{x\})$ is produced by replacing the literals x' and x by 1 in the original SOP formula. $\square$

We refer to the foregoing procedure, apparently first given by Mitchell [138], as the "replace-by-one trick."

Example 4.8.3 Let f be given, as in Example 4.8.2, by the formula

$$f = wx'y + v'w'x' + xz + xy' + vx'y'.$$

Then
$$
\begin{aligned}
EDIS(f, \{v\}) &= wx'y + w'x' + xz + xy' + x'y' \\
&= x' + y' + z \\
EDIS(f, \{w\}) &= x'y + v'x' + xz + xy' + vx'y' \\
&= x' + y' + z \\
EDIS(f, \{w, x\}) &= EDIS(EDIS(f, \{w\}), \{x\}) \\
&= 1 + y' + z = 1.
\end{aligned}
$$

$\square$

Example 4.8.4 We use the following (correct) calculations to illustrate potential pitfalls in applying the replace-by-1 trick:

(a) $\qquad\qquad EDIS(u' + vw, \{u\}) = 1 + vw = 1$
(b) $\qquad\qquad EDIS((u + v)', \{u\}) = EDIS(u'v', \{u\}) = v'$
(c) $\quad EDIS((u + v)(u' + w), \{u\}) = EDIS(uw + u'v + vw, \{u\})$
$\qquad\qquad\qquad\qquad\qquad\qquad\qquad = w + v.$

Calculation (a) illustrates that $EDIS(f, \{u\})$ is not found simply by deleting u' and u (which would produce vw rather than 1 in this case), but by replacing both u' and u by 1. Calculations (b) and (c) illustrate the need to express f in sum-of-products form before the literals u' and u are replaced by 1. If the replacements are made in the original formulas, the resulting erroneous calculations are:

(b') $\qquad\qquad EDIS((u + v)', \{u\}) = (1 + v)' = 0$
(c') $\quad EDIS((u + v)(u' + w), \{u\}) = (1 + v)(1 + w) = 1.$

$\square$

It is apparent from Theorems 4.8.2 and 4.8.3, and from the accompanying examples, that $ECON(f,T)$ tends to be "smaller" than f and that $EDIS(f,T)$ tends to be "bigger" than f. We formalize this observation as follows:

Theorem 4.8.4 *Let $f\colon \mathbf{B}^n \longrightarrow \mathbf{B}$ be a Boolean function expressed in terms of arguments $x, y, \ldots$ and let T be a subset of $\{x, y, \ldots\}$. Then*

$$ECON(f,T) \leq f \leq EDIS(f,T).$$

Proof. By Theorem 4.8.1, $ECON(f,T)$ may be expressed by a formula comprising only terms of $BCF(f)$. Expressed in such form, $ECON(f,T)$ is formally included (Sect. A.2) in $BCF(f)$; thus $ECON(f,T) \leq f$. To prove that $f \leq EDIS(f,T)$, we express f in the SOP form $f = \sum_i p_i$, whence $EDIS(f,T) = EDIS(\sum_i p_i, T) = \sum_i EDIS(p_i, T)$, the latter equality following from Theorem 4.8.3. It also follows from Theorem 4.8.3 that $p_i \leq EDIS(p_i, T)$ for all values of the index i. Hence

$$f = \sum_i p_i \leq \sum_i EDIS(p_i, T) = EDIS(f,T).$$

$\square$

Theorem 4.8.5 *Let f be an n-variable Boolean function, let U be a p-element subset of the argument-set $\{x_1, \ldots, x_n\}$, and let t be a q-argument term whose arguments are disjoint from those in U. Then*

$$
\begin{array}{llll}
\text{(a)} & ECON(f/t, U) & = & (ECON(f, U))\,/\,t \\
\text{(b)} & EDIS(f/t, U) & = & (EDIS(f, U))\,/\,t
\end{array}
$$

Proof. Let the arguments in $\{x_1, \ldots, x_n\}$ be ordered, without loss of generality, into blocks $X = \{T, U, V\}$ such that the term t comprises the arguments in T. Let K be the unique vector in $\{0,1\}^q$ such that $t = T^K$. Then identity (a) is proved by direct calculation:

$$ECON(f/t, U) = \prod_{A \in \{0,1\}^p} f(K, A, V) = \left(\prod_{A \in \{0,1\}^p} f(T, A, V) \right) / t$$

Identity (b) is proved by similar calculation, putting a Boolean sum in place of the foregoing product. $\square$

Theorem 4.8.6 *Let* $f: \mathbf{B}^n \longrightarrow \mathbf{B}$ *be a Boolean function expressed in terms of arguments* $x_1, x_2, \ldots, x_n$ *and let* S *be a subset of* $\{x_1, x_2, \ldots, x_n\}$. *Then*

$$(ECON(f, S))' = EDIS(f', S) \tag{4.39}$$
$$(EDIS(f, S))' = ECON(f', S). \tag{4.40}$$

Proof. We assume without loss of generality that $S = \{x_1, \ldots, x_m\}$, and define $T = \{x_{m+1}, \ldots, x_n\}$. Invoking Theorem 4.8.1 and De Morgan's laws,

$$
\begin{aligned}
(ECON(f(S, T), S))' &= (\ \prod_{A \in \{0,1\}^m} f(A, T)\ )' \\
&= \sum_{A \in \{0,1\}^m} f'(A, T) \\
&= EDIS(f'(A, T), S),
\end{aligned}
$$

verifying (4.39). Equation (4.40) is verified by dual computations. $\square$

Theorem 4.8.7 *Let* $f: \mathbf{B}^n \longrightarrow \mathbf{B}$ *be a Boolean function whose first* m *arguments are denoted* $x_1, x_2, \ldots, x_m$. *If the condition*

$$f(X, Y) = 0 \qquad (\forall X \in \mathbf{B}^m) \tag{4.41}$$

is satisfied, then so is the condition

$$EDIS(f, X) = 0. \tag{4.42}$$

Proof. Condition (4.41) implies that $f(A, Y) = 0$ for all A in $\{0, 1\}^m$; hence

$$\sum_{A \in \{0,1\}^m} f(A, Y) = 0,$$

from which (4.42) follows by Theorem 4.8.1. $\square$

4.9 Redundant Variables

An important problem in the practical application of Boolean algebra is to represent Boolean functions by formulas that are as simple as possible. One approach to the simplification of a Boolean formula is to minimize the number of variables appearing in it explicitly. This approach was investigated in 1938 by Shannon [183], who noted that a Boolean function f does not actually involve the variable x_k in case the condition

$$f/x_k' = f/x_k \tag{4.43}$$

holds identically. We say in this case that the variable x_k is *redundant* in f; the terms "vacuous" and "inessential" are also used to describe such a variable.

The functions of interest in Boolean reasoning typically occur as *intervals*, (*cf.* Section 2.4), *i.e.*, as sets of the form

$$[g, h] = \{f \mid g \le f \le h\}, \tag{4.44}$$

where $g, h\colon \mathbf{B}^n \longrightarrow \mathbf{B}$ are Boolean functions. Interval (4.44) is non-empty if and only if the condition $g \le h$ is satisfied. An "incompletely-specified" function (*cf.* Section 2.12), for example, is an interval of Boolean functions; as a further example, the set of solutions of a Boolean equation (Chapter 6) may be expressed as a system of intervals.

Let $X = \{x_1, \ldots, x_n\}$ denote the set of arguments of g and h, and let S be a subset of X. We say that S is a *redundancy subset* on an interval in case there is a function belonging to that interval in which all of the arguments in S are redundant. We say that S is a *maximal redundancy subset* on the interval in case (a) it is a redundancy subset on the interval and (b) S is not a proper subset of any redundancy subset on the interval.

The problem of finding maximal redundancy subsets has been investigated from a number of points of view: Hight [83] employs decomposition-charts [5]; Dietmeyer [50] applies array-operators; Kambayashi [98] reduces the location of such subsets to a covering problem; Halatsis and Gaitanis [76] generate a Boolean function whose prime implicants correspond to the maximal redundancy subsets; and Grinshpon [75] carries out a search-process aided by a numerical criterion. We approach the problem from yet another point of view in this section, based on the elimination-operators discussed in Section 4.8.

Theorem 4.9.1 *Let $g, h\colon \mathbf{B}^n \longrightarrow \mathbf{B}$ be Boolean functions expressed in terms of arguments $x_1, \ldots, x_n$, and let S be a subset of those arguments.*

1. *If f is a Boolean function in the interval $[g, h]$, and S is redundant in f, then f belongs to the interval*

$$[EDIS(g, S), \; ECON(h, S)]. \tag{4.45}$$

2. *If the interval* (4.45) *is non-empty, then there is a Boolean function in* $[g, h]$ *in which* S *is redundant.*

Proof.

1. (By induction on the number of elements in S.) Suppose f to be a member of $[g, h]$, whence the conditions

$$\begin{array}{ccccc} g/x'_k & \leq & f/x'_k & \leq & h/x'_k \\ g/x_k & \leq & f/x_k & \leq & h/x_k \end{array} \qquad (4.46)$$

hold for any argument x_k in $\{x_1, \ldots, x_n\}$. If the argument x_k is redundant in f, then the constraint

$$g/x'_k + g/x_k \ \leq \ f \ \leq \ h/x'_k \cdot h/x_k \qquad (4.47)$$

follows from (4.43) and (4.46), inasmuch as the redundancy of x_1 implies that $f(x_1, x_2, \ldots) = f(0, x_2, \ldots) = f(1, x_2, \ldots)$. Thus (4.45) is verified for the case in which S has one member. Assume next that the theorem holds if S has m members, *i.e.*, that if the variables $x_1, x_2, \ldots, x_m$ are redundant in f, then f belongs to the interval (4.45), where $S = \{x_1, x_2, \ldots, x_m\}$. If the variable x_{m+1} is also redundant in f, then f belongs by the induction hypothesis to the interval

$$[EDIS(EDIS(g, S), \{x_{m+1}\}) \ ECON(ECON(h, S), \{x_{m+1}\})] \,. \qquad (4.48)$$

Recalling the definition of $EDIS$ and $ECON$, however, we may express interval (4.48) as follows:

$$[EDIS(g, \ S \cup \{x_{m+1}\}), \ ECON(h, \ S \cup \{x_{m+1}\})] \,. \qquad (4.49)$$

Thus the theorem holds if S has $m + 1$ members, proving Part 1 of the theorem.

2. It follows from Theorem 4.8.4 that interval (4.45) is a subset of $[g, h]$. If (4.45) is non-empty, then $EDIS(g, S)$ is a function belonging to $[g, h]$ in which S is redundant, proving Part 2 of the theorem.

$\square$

We define the *resultant of removal* of variable x from interval $[p, q]$ to be the interval $[EDIS(p, \{x\}), \ ECON(q, \{x\})]$. As noted in the proof of Theorem 4.9.1, the resultant of removal of a variable from an interval is a subset of that interval. It should be noted that the resultant of removal of x from $[p, q]$ is different from the resultant of *elimination* of x from $[p, q]$ The latter resultant takes the form $[ECON(p, \{x\}), EDIS(q, \{x\})]$ (the proof is assigned as an exercise), which is a superset of the interval $[p, q]$.

Theorem 4.9.1 shows that the maximal redundancy subsets on an interval $[g, h]$ may be determined by a tree-search through a space of intervals derived from $[g, h]$. The root of the tree is $[g, h]$; each child-node of a node $[p, q]$ is the resultant of removal of a variable not yet removed in the path leading to $[p, q]$. A variable is removed from an interval (and thus the search proceeds beyond the corresponding node) only if the resultant of removal of that variable is non-empty, *i.e.*, only if that variable is redundant on that interval. If no variable is redundant on a given interval in the search-space, then the variables removed in the path leading to that interval constitute a maximal redundancy subset.

The efficiency of the search-process clearly depends on the efficiency with which the redundancy of a given variable on a given interval can be decided. It follows from Theorem 4.9.1 that variable x is redundant on $[p, q]$ if and only if the condition

$$EDIS(p, \{x\}) \leq ECON(q, \{x\}) \tag{4.50}$$

is satisfied. If p is expressed in arbitrary sum-of-products form and q is expressed in Blake canonical form, then

- $EDIS(p, \{x\})$ is found by deleting x and x' wherever they occur in a term (if either x or x' appears alone as a term, then $EDIS(p, \{x\}) = 1$);

- $ECON(q, \{x\})$ is found by deleting any term in q that contains either x or x' (the result remains in Blake canonical form); and

- condition (4.50) is satisfied if and only if each term of $EDIS(p, \{x\})$ is included in some term of $ECON(q, \{x\})$.

The termwise comparison described above suffices as a test for inclusion because $ECON(q, \{x\})$ is expressed in Blake canonical form. A formula whose included sum-of-products formulas may be tested by termwise comparison is called *syllogistic*. Syllogistic formulas are discussed in Appendix A, where it is demonstrated that a Blake canonical form is syllogistic.

We call a subset T of X a *minimal determining subset* on the interval $[g, h]$ provided T has the following properties:

1. the variables in T suffice to describe at least one function in $[g, h]$, and

2. no proper subset of T has property 1.

Each minimal determining subset, T, on $[g, h]$ (and nothing else) is the relative complement with respect to X of a maximal redundancy subset, S, on $[g, h]$ i.e., $T = X - S$.

Example 4.9.1 Let us determine the minimal determining subsets on the interval $[g, h]$, where g and h are given by the formulas

$$g = v'w'xy'z + vw'x'yz' \tag{4.51}$$
$$h = v'x + vx' + w' + y + z . \tag{4.52}$$

A depth-first search through the space of intervals derived from $[g, h]$ by variable-removals is indicated in Table 4.1. The maximal redundancy subsets found in this search are $\{v, w, x\}$, $\{v, x, y, z\}$, and $\{w, y, z\}$. The corresponding minimal determining subsets are, respectively, $\{y, z\}$, $\{w\}$, and $\{v, x\}$. The function-intervals associated with these subsets are shown in Table 4.2. □

Example 4.9.2 (Grinshpon [75])   An incompletely-specified switching function $f: \{0, 1\}^6 \longrightarrow \{0, 1\}$ is described by the statements

$$\phi_0(X) = 1 \implies f(X) = 0$$
$$\phi_1(X) = 1 \implies f(X) = 1 , \tag{4.53}$$

the functions ϕ_0 and ϕ_1 being expressed as follows:

$$\phi_0 = x_6'[x_1'x_2'x_3' + x_5'(x_1'(x_2' + x_3' + x_4') + x_2'x_3')]$$
$$\phi_1 = x_1x_2[x_3x_5x_6' + x_6(x_3' + x_4' + x_5')] + x_5x_6(x_1x_2'x_4 + x_1'x_3) . \tag{4.54}$$

Thus f is any function in the interval $[g, h]$, where

$$g = \phi_1$$
$$h = \phi_0' . \tag{4.55}$$

Repeating the search-process described in Example 4.9.1 to find the maximal redundancy subsets, and computing complements of those subsets relative to $\{x_1, x_2, x_3, x_4, x_5, x_6\}$, we derive the following minimal determining subsets: $\{x_1, x_3, x_6\}$, $\{x_2, x_3, x_4\}$, $\{x_2, x_5, x_6\}$, $\{x_3, x_5, x_6\}$, $\{x_1, x_2, x_3, x_5\}$, and $\{x_1, x_2, x_6\}$. □

Subset	Test			Redundant?
$\emptyset$	$v'w'xy'z + vw'x'yz'$	$\leq$	$v'x + vx' + w' + y + z$	yes
$\{v\}$	$w'xy'z + w'x'yz'$	$\leq$	$w' + y + z$	yes
$\{v, w\}$	$xy'z + x'yz'$	$\leq$	$y + z$	yes
$\{v, w, x\}$	$y'z + yz'$	$\leq$	$y + z$	yes
$\{v, w, x, y\}$	$z + z'$	$\leq$	z	no
$\{v, w, x, z\}$	$y' + y$	$\leq$	y	no
$\{v, x\}$	$w'y'z + w'yz'$	$\leq$	$w' + y + z$	yes
$\{v, x, y\}$	$w'z + w'z'$	$\leq$	$w' + z$	yes
$\{v, x, y, z\}$	w'	$\leq$	w'	yes
$\{w\}$	$v'xy'z + vx'yz'$	$\leq$	$v'x + vx' + y + z$	yes
$\{w, x\}$	$v'y'z + vyz'$	$\leq$	$y + z$	yes
$\{w, x, y\}$	$v'z + vz'$	$\leq$	z	no
$\{w, x, z\}$	$v'y' + vy$	$\leq$	y	no
$\{w, y\}$	$v'xz + vx'z'$	$\leq$	$v'x + vx' + z$	yes
$\{w, y, z\}$	$v'x + vx'$	$\leq$	$v'x + vx'$	yes

Table 4.1: Development of maximal redundancy subsets.

Minimal Determining Subset	Function-Interval
$\{y, z\}$	$[y'z + yz',\ y + z]$
$\{w\}$	$[w',\ w']$
$\{v, x\}$	$[v'x + vx',\ v'x + vx']$

Table 4.2: Minimal determining subsets and associated intervals.

4.10 Substitution

We have seen that a variable may be removed from a Boolean formula by calculating one of the following with respect to that variable:

- its quotient,
- its conjunctive eliminant,
- its disjunctive eliminant, or
- its Boolean derivative.

Another way to remove a variable is by *substitution.* Suppose we are given the formula

$$a'xy + bx'z \tag{4.56}$$

and we wish to remove x by the substitution

$$x = cy. \tag{4.57}$$

If we replace each appearance of x in the original formula by the formula cy, and each appearance of x' by $c' + y'$, the result after simplification is

$$a'cy + bc'z + by'z. \tag{4.58}$$

Such direct replacement is natural for hand-calculation, but can be awkward to automate. A more readily-automated procedure is provided by the following theorem.

Theorem 4.10.1 *Let f and g be Boolean formulas on a common Boolean algebra $\mathbf{B}$, and let x be one of the variables appearing in the formula f. The result of substituting g for x in the formula f is given by either of the following formulas:*

$$ECON(f + (x \oplus g), \{x\}) \tag{4.59}$$

$$EDIS(f \cdot (x' \oplus g), \{x\}). \tag{4.60}$$

Proof. The result of substituting $x = g$ into the formula $f(x, y, \ldots)$ is the formula $f(g, y, \ldots)$, which has the expanded form

$$[f(0, y, \ldots) + g] [f(1, y, \ldots) + g'].$$

The latter formula may be expressed equivalently as

$$ECON(x'(f(0, y, \ldots) + g) + x(f(1, y, \ldots) + g'), \{x\}),$$

which is equivalent in turn to (4.59). Similar calculations verify (4.60). $\square$

Example 4.10.1 Applying (4.59), the result of substituting $x = cy$ in formula (4.56) is the formula

$$ECON(a'xy + bx'z + (x \oplus cy), \{x\}),$$

i.e.,

$$[bz + cy][a'y + c' + y'],$$

which takes the form (4.58) when multiplied out and simplified. □

Example 4.10.2 The circuit of Example 4.7.2 is described by the equations

$$z_1 = x_1 x_2 \tag{4.61}$$
$$z_2 = x_3 + z_1 . \tag{4.62}$$

The consequent

$$z_2 = x_3 + x_1 x_2 \tag{4.63}$$

was obtained in that example by eliminating x_1. An alternative approach is to perform the substitution (4.61) in the right-hand side of equation (4.62). Applying (4.59), we write

$$z_2 = ECON(x_3 + z_1 + (z_1 \oplus x_1 x_2), \{z_1\})$$

which yields (4.63). If we apply (4.60), *viz.*,

$$z_2 = EDIS((x_3 + z_1) \cdot (z_1' \oplus x_1 x_2), \{z_1\}),$$

we obtain the same result. □

Example 4.10.3 Given a Boolean function f and a term t, the Boolean quotient f/t (Section 2.15) is the function that results when the substitution $t = 1$ is carried out in f. Thus

$$f/t = ECON(f + (t \oplus 1), T)$$
$$= ECON(f + t', T)$$

where T is the set of variables in the term t. Let us apply formal substitution to evaluate the quotient $f(x, y, z)/xy'$:

$$f(x, y, z)/xy' = ECON(f + x' + y, \{x, y\})$$
$$= (f(0, 0, z) + 1)(f(0, 1, z) + 1)(f(1, 0, z) + 0)(f(1, 1, z) + 1)$$
$$= f(1, 0, z).$$

Thus formal substitution produces the same result as do the methods of Section 2.15. □

4.11 The Tautology Problem

A basic problem in propositional logic and Boolean reasoning is to decide whether the members of a given set of terms (products) sum to one. Specifically, the problem is to determine whether the relation

$$t_1(X) + t_2(X) + \ldots + t_m(X) = 1 \qquad (4.64)$$

is an identity, where $t_1, \ldots, t_m$ are products of variables (complemented or uncomplemented) from the set $X = \{x_1, \ldots, x_n\}$.

This problem arises in proving theorems in the propositional calculus [47, 51, 53], in deciding the consistency (*i.e.*, solvability) of a Boolean equation [221] and in determining minimal formulas for switching functions [18, 37, 70, 175, 176]. See Galil [62] for a detailed study of the complexity of this problem.

4.11.1 Testing for Tautology

A boolean formula is a tautology, clearly, if evaluating its Blake canonical form produces the 1-formula. More efficient tests [175, 51, 47, 222] are based on the fact that a Boolean formula F is a tautology if and only if its Boolean quotients with respect to x' and x are tautologies, where x is any one of its arguments. Each such test employs an elaboration of the following rules:

1. If F is is empty (*i.e.*, it contains no terms), then F is not a tautology.

2. If F contains the term 1, then F is a tautology.

3. Otherwise, F is a tautology if and only if F/u' is a tautology and F/u is a tautology, where u is an argument of F.

Example 4.11.1 To improve efficiency in hand-calculation, we amend the second of the foregoing rules to read as follows: "If F contains the term 1, or a pair u and u' of single-letter terms, then F is a tautology." Let us apply the amended rule-set to the formula

$$F = w'y'z + xy + yz + x'z' + w'x + wy' .$$

F does not satisfy Rule 1 or Rule 2; therefore we evaluate F/x' and F/x' (variable x is chosen arbitrarily):

$$F/x' = w'y'z + yz + z' + wy'$$
$$F/x = w'y'z + y + yz + w' + wy' .$$

F is a tautology if and only if each of the foregoing quotients is a tautology. Neither quotient satisfies Rule 1 or Rule 2; therefore we divide each by some letter and its complement:

$$
\begin{aligned}
(F/x')/y' &= F/x'y' &= w'z + z' + w \\
(F/x')/y &= F/x'y &= z + z' \\
(F/x)/y' &= F/xy' &= w'z + w' + w \\
(F/x)/y &= F/xy &= 1 + z + w'
\end{aligned}
$$

The only one of the foregoing quotients not verified to be a tautology by either Rule 1 or the amended Rule 2 is $(F/x')/y' = F/x'y' = w'z + z' + w$. We therefore generate quotients with respect either to w or to z; we choose w:

$$
\begin{aligned}
(F/x'y')/w' &= F/x'y'w' &= z + z' \\
(F/x'y')/w &= F/x'y'w &= z' + 1
\end{aligned}
$$

Every formula has thus been verified by one of the rules to be a tautology; hence F is a tautology. $\square$

4.11.2   The Sum-to-One Theorem

It is frequently necessary in Boolean calculations to determine if a given term is included in a given Boolean function. The following theorem, employed by Samson & Mueller [175] and Ghazala [70] to simplify switching formulas, transforms the problem of determining such inclusion into one of determining the tautology of an associated function.

Theorem 4.11.1 *Let f be a Boolean function and let t be a term. Then*

$$
t \leq f \quad \Longleftrightarrow \quad f/t = 1 . \tag{4.65}
$$

Proof.   The implication $\Longrightarrow$ follows from Proposition 2.15.1 for $f \mapsto t$ and $g \mapsto f$, while the converse implication follows from Proposition 2.15.4. $\square$

Example 4.11.2 Let $f = xy' + x'z + yz'$ and let $t = y'z$. To decide if $t \leq f$, we evaluate f/t:

$$
(xy' + x'z + yz')/y'z = x + x' = 1 .
$$

Thus $y'z \leq xy' + x'z + yz'$. $\square$

4.11.3 Nearly-Minimal SOP Formulas

A much-studied problem in switching theory is to find minimal SOP formulas for Boolean functions. This problem has little direct importance in Boolean reasoning; for computational efficiency, however, it is useful to be able to represent a function by a nearly-minimal formula. The sum-to-one theorem enables this to be done conveniently for a function expressed (as is usual in reasoning-computations) in Blake canonical form.

It was shown by Quine [161] that the terms of a least-cost SOP formula for a Boolean function f are necessarily prime implicants of f—provided that the cost of a formula increases if the number of literals in a term increases. We assume such a cost-measure; thus a simplified SOP formula for f is a subformula of $BCF(f)$. An *irredundant* formula for f is a disjunction of prime implicants of f that (a) represents f and (b) ceases to represent f if any of its terms is deleted. The search for simplified SOP formulas therefore need only be over the irredundant formulas.

The following procedure converts $BCF(f)$ into an irredundant formula for f by a succession of term-deletions. The cost of an SOP formula is assumed to be the total number of literals in the formula. The procedure attempts to minimize this cost by considering the most expensive terms (those comprising the most literals) first for deletion. This procedure does not guarantee minimality, but typically produces minimal or near-minimal costs.

Step 1. Sort the terms of $BCF(f)$ according to the number of literals they contain, putting those having the most literals first. Denote by F the resulting formula.

Step 2. Let T be a term of F and let $F - T$ denote the formula that results when T is removed from F. Beginning with the first term in F, carry out the following process until all terms T have been considered:

If $(F - T)/T$ is a tautology, replace F with $F - T$; otherwise do nothing.

Return the resulting formula.

Step 1 generates all of the candidate-terms for an irredundant formula, and arranges that the most costly terms (those having the largest number of literals) will be considered first for removal. Step 2 makes use of the sum-to-one theorem (Theorem 4.11.1) to produce an irredundant representation F for f; no term of F is included in the remainder of F.

Example 4.11.3 To produce a nearly-minimal formula corresponding to the formula

$$f = ABE' + CD'E + AC'D'E' + ABDE + A'B'CD + AB'C + A'B'C'D' ,$$

we first carry out Step 1 of the foregoing procedure, *i.e.*, we generate the terms of $BCF(f)$ and arrange them in descending order of the number of literals they contain:

$$\begin{aligned} BCF(f) \quad &= \quad B'C'D'E' + A'B'D'E + A'B'C'D' + ABE' + AD'E' + \\ &= \quad +B'CE + CD'E + ABD + B'CD + AC . \end{aligned}$$

The formula remaining after completion of Step 2 is

$$f = CD'E + ABD + B'CD + A'B'C'D' + AD'E' . \qquad (4.66)$$

It can be shown that there are four irredundant formulas for f; each has the form

$$f = CD'E + ABD + B'CD + < \text{OTHER TERMS} >$$

where $<$ OTHER TERMS $>$ may be one of the following:

$$\begin{array}{llll} A'B'C'D' & + & AD'E' & \\ B'C'D'E' & + & A'B'D'E & + & AD'E' \\ B'C'D'E' & + & A'B'C'D' & + & ABE' & + & AC \\ B'C'D'E' & + & A'B'D'E & + & ABE' & + & AC \end{array}$$

Thus 4.66 is a least-cost formula for the given function. □

Exercises

1. Prove or disprove: any two inconsistent Boolean systems are equivalent.

2. (Couturat [41, p. 36]) Prove the following equivalence:
$$
\begin{cases}
a & \leq & b \oplus c \\
b & \leq & a \oplus c \\
c & \leq & a \oplus b
\end{cases}
\iff
\begin{cases}
a & = & b \oplus c \\
b & = & a \oplus c \\
c & = & a \oplus b
\end{cases}
$$

3. Prove that the following implication is valid:
$$
\begin{cases}
a \leq & x & \leq b \\
f(a) & = & f(b)
\end{cases}
\implies f(x) = f(a)
$$

4. (Löwenheim [124], Rudeanu [172]). Let f, g, and h be Boolean functions and assume that $f(X) = 0$ is consistent. Show that
$$
[f(X) = 0 \implies g(X) = h(X)] \iff [gf' = hf'] .
$$

5. Let f, g, and h be Boolean functions. Show that
$$
[f(X) = 0 \implies g(X) = 0] \implies
$$
$$
[\,[g(X) = 0 \implies h(X) = 0] \implies [f(X) = 0 \implies h(X) = 0]\,]
$$

6. (Rudeanu [172, p. 100]). Show that $S \implies [T \iff U]$ is valid, where S, T, and U are defined as follows:

   S: $x \leq a + b'$ and $y \leq a' + b$

   T: $b'x + ay' = a$ and $bx' + a'y = b$

   U: $x \leq ab'$ and $y \leq a'b$

7. The RS-latch, a basic component in digital circuits, is characterized by the coupled equations
$$
Q = S + X' \tag{4.67}
$$
$$
X = R + Q' , \tag{4.68}
$$
where R and S are the latch's inputs and Q and X are its outputs. For proper operation, the inputs should be constrained to satisfy the

condition $RS = 0$, in which case we wish to verify that the outputs will satisfy the condition $X = Q'$. The problem: show that the system (4.67, 4.68) implies the condition

$$RS = 0 \quad \Longrightarrow \quad X = Q'.$$

8. In the earliest published work applying Boolean algebra to switching circuits, Nakasima [146] lists rules, shown below, to assist in the solution of Boolean equations. Verify each rule.

(a) If $A + X = A$, then $X \leq A$.

(b) If $AX = A$, then $A \leq X$.

(c) If $A + X = B$, then $A'B \leq X \leq A + B$.

(d) If $AX = B$, then $AB \leq X \leq A' + B$.

(e) If $AX + B = C$, then $B'C \leq AX \leq B + C$
and $(A + B)C \leq B + X \leq A'B' + C$.

(f) If $(A + X)B = C$, then $BC \leq A + X \leq B' + C$
and $(A' + B')C \leq BX \leq AB + C$.

(g) If $A + X = B$ and $A' + X = C$, then $X = BC$.

(h) If $AX = B$ and $A'X = C$, then $X = B + C$.

(i) If $A + X = B$ and $AX = C$, then $X = B(A' + C)$.

9. Assume that $B = \{0, 1, a', a\}$. It is asserted in the text that:

(a) the set of solutions of the equation $a'x + y = 0$ is $\{(0,0), (a,0)\}$; and

(b) if $f(x, y) = 0$ has solutions $(0, 0)$ and $(a, 1)$ then it must also have solutions $(0, a')$ and (a, a).

Prove these assertions.

10. Given the Boolean system

$$awx' + bx = b'x'y'$$
$$by' + a \leq x + y,$$

(a) Reduce to a single equivalent equation of equals-zero form.

(b) Reduce to a single equivalent equation of equals-one form.

(c) Eliminate x, expressing the resultant in equals-one form.

Express all resulting formulas in Blake canonical form.

11. Let $B = \{0, 1, a', a\}$ and define a constraint on (x, y) as follows:

$$(x, y) \in \{(0, a), (1, a'), (a', 0)\} \ .$$

Decide whether this is a Boolean constraint, making your reasoning explicit.

12. A constraint on the variables x, y, z is expressed by the Boolean equation

$$ax'y + bz = a'yz + xz' \ .$$

Express the same constraint by an equation of the form

$$g(x, y, z) = a + z \ ,$$

representing the function g in Blake canonical form.

13. The equation $ab + x = b'$ is equivalent to the equation $f(x) = 0$ and also to the equation $g = a + b$. Express the functions f and g by simplified formulas.

14. Given

$$\begin{aligned} f \ = \ & AE'F' + DEF + BDE' + A'B'E'F' + BCD'F+ \\ & +BDEF' + B'D'EF + BD'E'F' \ , \end{aligned}$$

(a) Express f in Blake canonical form.

(b) Assume the resultant of elimination of the variables B and C from the equation $f = 0$ to be the equation $g = 0$. Express the function g in simplified form. Explain your method.

15. The system

$$\begin{aligned} ax + y \ &= \ xy \\ x' + y \ &\leq \ a + y \end{aligned}$$

expresses a constraint on the variables x and y over the Boolean algebra $B = \{0, 1, a', a\}$. Express the same constraint by the equation $ax = g(x, y)$, writing $g(x, y)$ as a simplified formula.

16. Let $B = \{0, 1, a', a\}$ and define constraints on (x, y) as follows:

    (a) $(x, y) \in \{(a, a), (1, 1)\}$
    (b) $(x, y) \in \{(a, a), (a, 1), (1, 1)\}$

Decide in each case whether the constraint is Boolean, making your reasoning explicit.

17. Given

$$f = AE'F' + DEF + BDE' + A'B'E'F' + BCD'F + \\ + BDEF' + B'D'EF + BD'E'F',$$

    (a) Express f in Blake canonical form.
    (b) Assume the resultant of elimination of the variables B and C from the equation $f = 0$ to be the equation $g = 0$. Express the function g in simplified form. Explain your method.

18. Let S be a subset of the arguments in formulas representing Boolean functions g and h. Show that the resultant of elimination of the arguments in S from the interval

$$g \leq x \leq h$$

is the interval

$$ECON(g, S) \leq x \leq EDIS(h, S).$$

19. Show that the conditions

$$f(0, x_2, \ldots) = f(1, x_2, \ldots) \tag{4.69}$$

and

$$EDIS(f, x_1) = ECON(f, x_1) \tag{4.70}$$

are equivalent.

20. Derive the minimal determining subsets listed in Example 4.9.2 by application of the search-procedure described in Section 4.9.

21. Devise a procedure to convert a given SOP formula into an equivalent *orthogonal* SOP formula having the fewest possible terms (*cf.* Section 2.13).

Chapter 5

Syllogistic Reasoning

We outline in this chapter an approach, which we call "syllogistic," to the solution of logical problems. The essential features of the syllogistic approach were formulated by Blake [10].

The examples we consider are expressed either in the algebra of propositions or the algebra of classes; these algebras are discussed in Chapter 2. The elementary units of reasoning in class-logic are classes, i.e., subsets of a universal set. George Boole's "algebra of logic" [12, 13], for example, was formulated in terms of classes. The elementary units in propositional logic, on the other hand, are propositions, i.e., statements that are necessarily either true or false. Although the examples we present are in terms either of propositions or classes, the methods we discuss are applicable in any Boolean algebra.

Syllogistic reasoning makes use of just one rule of inference, rather than the many rules conventionally employed. It proceeds by applying the operation of consensus repeatedly to a formula f representing given logical data. We have seen (Chapter 3) that such repeated consensus-generation produces $BCF(f)$, the Blake canonical form for f, together possibly with additional terms that are absorbed by the terms in $BCF(f)$. Syllogistic reasoning is thus intimately associated with the Blake canonical form.

Syllogistic reasoning is related to the resolution-based techniques employed in predicate logic. The terms of syllogistic formulas are the duals, in the Boolean domain, of the clauses of predicate logic. The operation of consensus is the Boolean dual of resolution. Syllogistic reasoning differs from reasoning in predicate logic, however, in one important way. Resolution in the predicate calculus is employed as part of a strategy of theorem-proving by refutation; a problem is formulated as a theorem, which is proved (possibly assigning values to variables as a side-effect) by conjoining the theorem's denial with the premises and deducing a contradiction. In syllogistic reasoning, on the other hand, the strategy is to chain forward from the premises, represented by an equation of the form $f = 0$, until all of the "prime" consequents are generated. A given consequent (theorem) can then be verified from the prime consequents by simple term-by-term comparisons.

Forward chaining is not feasible in predicate logic because it is not guaranteed to terminate. Even in the finite Boolean domain (where termination is guaranteed), theorem-proving typically requires more computation by forward chaining than by refutation. In most applications, however, Boolean problems are not formulated as theorems to be proved. Forward chaining is typically the first step in the solution of such problems by Boolean reasoning, after which appropriate operations of Boolean analysis (e.g., elimination, division, substitution, solution) are performed.

Example 5.0.1 Suppose that our knowledge concerning the endeavors of a certain college student is expressed by the following statements [97]:

1. If Alfred studies, then he receives good grades.

2. If Alfred doesn't study, then he enjoys college.

3. If Alfred doesn't receive good grades, then he doesn't enjoy college.

What may we conclude concerning Alfred's academic performance?

A correct (but probably not obvious!) conclusion is that Alfred receives good grades. Our object in this chapter is to show how syllogistic reasoning enables us to arrive at such a conclusion mechanically.

5.1   The Principle of Assertion

Information is conveyed in ordinary algebra by equations. Boole and other nineteenth-century logicians therefore found it natural to write logical statements as equations. To analyze a collection of statements in Boole's algebra of logic, the corresponding equations are reduced to a single equivalent equation of the form

$$f(A, B, C, ...) = 0 ,\qquad(5.1)$$

where f is a Boolean function and $A, B, C, \ldots$ are symbols which, in Boole's formulation, represent classes of objects.

All of the properties of Boolean equations remain valid if the symbols $A, B, C, \ldots$ in (5.1) are propositions, rather than classes, and if 0 and 1 represent, respectively, the identically false and identically true propositions. Certain statements not involving equations, however, are valid only for propositions. These statements derive from an axiom peculiar to the calculus of propositions, called the *principle of assertion* (see Couturat [41]), which may be stated as follows:

$$[A = 1] = A . \tag{5.2}$$

In Couturat's words, "To say that a proposition A is true is to state the proposition itself." It is therefore possible in the calculus of propositions to dispense entirely with equations. If $f(A, B, C, \ldots)$ is a propositional (i.e., two-valued) function, then equation (5.1) may be stated equivalently by the proposition

$$f'(A, B, C, \ldots) . \tag{5.3}$$

Modern logicians have abandoned equations in the formulation of propositional logic. We shall employ the classical equation-based approach, however, in order to apply the techniques of Boolean analysis without modification to problems in propositional logic.

Let us convert the set of premises in Example 5.0.1, concerning Alfred's collegiate endeavors, to a single equivalent equation of the form (5.1). We begin by expressing the three given premises by the system

$$
\begin{array}{cccc}
1. & S & \longrightarrow & G \\
2. & S' & \longrightarrow & E \\
3. & G' & \longrightarrow & E'
\end{array}
\tag{5.4}
$$

of propositions, the symbols E, G, and S being defined as follows:

$$
\begin{array}{rcl}
E & = & \text{"Alfred enjoys college"} \\
G & = & \text{"Alfred gets good grades"} \\
S & = & \text{"Alfred studies".}
\end{array}
$$

The premises in (5.4) may be represented equivalently by the system

$$
\begin{array}{rcl}
SG' & = & 0 \\
S'E' & = & 0 \\
G'E & = & 0 ,
\end{array}
\tag{5.5}
$$

of propositional equations. We thus arrive at a single equation equivalent to
the set of premises given in Example 5.0.1, i.e.,

$$SG' + S'E' + G'E = 0 .$$

$$(5.6)$$

5.2 Deduction by Consensus

In traditional logic, deduction is carried out by invoking a number of rules of
inference; these rules announce that certain conclusions follow from certain
sets of premises. Some logic-texts, e.g., [97], list hundreds of such rules.
A cardinal advantage of syllogistic reasoning is that it employs only one
rule of inference, that of hypothetical syllogism. This rule states that the
conclusion given below follows from its premises (we express the components
of this syllogism both as conditionals and as equations):

	Conditional	Equation
Major Premise	$A \longrightarrow B$	$AB' = 0$
Minor Premise	$B \longrightarrow C$	$BC' = 0$
Conclusion	$A \longrightarrow C$	$AC' = 0$

The two premises may be expressed by the single equation $f = 0$, where
f is given by

$$f = AB' + BC' .$$

$$(5.7)$$

The conclusion is expressed by the equation $g = 0$, where

$$g = AC'.$$

$$(5.8)$$

The problem of deduction in this case is that of obtaining the term AC'
(representing the conclusion) from the terms AB' and BC' (representing
the premises). We note that AC' is the consensus of the terms AB' and
BC'; thus, reasoning by the rule of hypothetical syllogism is carried out by
producing the consensus (which Blake called the "syllogistic result") of the
terms representing the premises. The utility of the consensus-operation is not
confined, however, to simple syllogisms. The single operation of consensus
suffices to produce a simple representation of all conclusions to be inferred
from any set of premises in propositional logic. Repeated application of
consensus to an SOP formula f, followed by absorption, produces $BCF(f)$,
i.e., the disjunction of all of the prime implicants of f. Thus, given a set of
premises reducible to an equation of the form (5.1) , the formula $BCF(f)$

represents (in a way we shall subsequently make precise) all of the conclusions that may be inferred from those premises.

Let us continue our study of Alfred's collegiate endeavors. Converting the left side of (5.6) to Blake canonical form, we represent everything we know about Alfred by the equation

$$S'E' + G' = 0 . \tag{5.9}$$

Equation (5.9) is equivalent to the system

$$S'E' \;=\; 0 \tag{5.10}$$
$$G' \;=\; 0 , \tag{5.11}$$

whose components may be given the verbal interpretations

(a) "Alfred studies or Alfred enjoys college."
(b) "Alfred gets good grades."

Statement (a) is a re-phrasing of one of the original premises; statement (b) is the not-very-obvious conclusion announced at the beginning of this chapter.

5.3 Syllogistic Formulas

Let $A, B, C, \ldots$ be Boolean variables and suppose that we are given a system of statements (premises) reducible to the equation (5.1). Let us suppose further that (5.1) is consistent. A *consequent* (or *conclusion*) of (5.1) is a statement or system of statements reducible to the equation

$$g(A, B, C, \ldots) = 0 \tag{5.12}$$

such that the implication

$$f = 0 \quad \Longrightarrow \quad g = 0 \tag{5.13}$$

is satisfied. Thus, by Theorem 4.4.1, (the Extended Verification Theorem), equation (5.12) is a consequent of equation (5.1) if and only if the relation

$$g \leq f \tag{5.14}$$

holds. Looking for consequents of the equation $f = 0$ is equivalent, therefore, to looking for functions g included in f. Let p be a prime implicant of f.

Then we call the equation $p = 0$, which is clearly a consequent of $f = 0$, a *prime consequent* of $f = 0$.

We assume henceforth that all Boolean functions are expressed by SOP (sum-of-products) formulas. Deciding whether a given SOP formula is included in another is not in general an easy task. It is not obvious, for example, that the function

$$g = BC'D + AD' \tag{5.15}$$

is included in the function

$$f = AC' + CD' + A'D. \tag{5.16}$$

Comparing two SOP formulas for inclusion becomes much easier, however, if we confine ourselves to *formal inclusion*. Recalling the the definition given in Chapter 3, we say that g is *formally included* in f, written $g \ll f$, in case each term of g is included in (i.e., has all the literals of) some term in f. It is clear that formal inclusion implies inclusion; that is,

$$g \ll f \quad \Longrightarrow \quad g \leq f \tag{5.17}$$

for all SOP formulas f and g. We call an SOP formula f *syllogistic* in case the converse of (5.17) also holds, i.e., in case the implication

$$g \leq f \quad \Longrightarrow \quad g \ll f \tag{5.18}$$

holds for all Boolean formulas g. Thus *an SOP formula is syllogistic if and only if every SOP formula that is included in it is also formally included in it.*

It is shown in Appendix A (Theorem A.3.1) that an SOP formula for a Boolean function f is a syllogistic representation of f if and only if it contains all the prime implicants of f. It follows directly that the simplest syllogistic formula for f is $BCF(f)$.

Suppose that a Boolean function f is expressed by a syllogistic formula, e.g., by $BCF(f)$. Then one may tell by inspection (or conveniently program a computer to tell) if any given SOP formula is included in f. Consider for example the function defined by (5.16); in Blake canonical form,

$$BCF(f) = AC' + CD' + A'D + AD' + C'D + A'C . \tag{5.19}$$

Deciding whether the function g defined by (5.15) is included in (5.19), unlike deciding whether g is included in the equivalent formula (5.16), is a simple matter of inspection. The terms $BC'D$ and AD' of (5.15) are included, respectively, in the terms $C'D$ and AD' of $BCF(f)$; hence, g is formally included in (5.19), and therefore included in (5.16).

5.4 Clausal Form

Suppose a Boolean function f is expressed as an SOP formula, i.e.,

$$f = p_1 + p_2 + \ldots + p_k, \tag{5.20}$$

where $p_1, \ldots, p_k$ are terms (products). Then the equation $f = 0$ is equivalent
to the system

$$
\begin{aligned}
p_1 &= 0 \\
p_2 &= 0 \\
&\;\vdots \\
p_k &= 0 .
\end{aligned}
\tag{5.21}
$$

Consider any term

$$p_i = a_1 \cdots a_m b'_1 \cdots b'_n$$

of (5.20). (If $m = 0$, i.e., if no a's are present, we consider p_i to have the
form $1 \cdot b'_1 \cdots b'_n$; if $n = 0$, we consider p_i to have the form $a_1 \cdots a_m \cdot 0'$.) The
equation $p_i = 0$ may be written in the equivalent form

$$a_1 \cdots a_m \cdot b'_1 \cdots b'_n = 0 . \tag{5.22}$$

We call a statement having the specialized form (5.22) a *clause*, and say
that it is in *clausal form*. A clause whose left-hand side is a prime implicant
of f will be called a *prime clause* of $f = 0$. If the a's and b's are propositions,
then we may write the clause (5.22) equivalently as a conditional, i.e.,

$$a_1 \cdots a_m \longrightarrow b_1 + \cdots + b_n \tag{5.23}$$

which we also call a clause and which we read as

$$\text{``IF } a_1 \text{ AND } \cdots \text{ AND } a_m, \text{ THEN } b_1 \text{ OR } \cdots \text{ OR } b_n \text{''.} \tag{5.24}$$

If $m = 0$, then (5.24) degenerates to $1 \longrightarrow b_1 + \cdots + b_n$, which may be read

$$\text{``}b_1 \text{ OR } \cdots \text{ OR } b_n\text{''.}$$

If $n = 0$, then (5.24) degenerates to $a_1 \cdots a_m \longrightarrow 0$, for which a direct (if
awkward) reading is

$$\text{``IT IS NOT THE CASE THAT } a_1 \text{ AND } \cdots \text{ AND } a_m \text{''.}$$

Some examples of equations of the form $p_i = 0$, together with their corresponding clauses, are tabulated below.

Equation		Clause	
$AB'CD'E'$	$= 0$	$AC \longrightarrow$	$B + D + E$
abx	$= 0$	$abx \longrightarrow$	0
$U'V'W'$	$= 0$	$1 \longrightarrow$	$U + V + W$

Example 5.4.1 Alice, Ben, Charlie, and Diane are considering going to a Halloween party. The social constraints governing their attendance are as follows:

1. If Alice goes then Ben won't go and Charlie will.

2. If Ben and Diane go, then either Alice or Charlie (but not both) will go.

3. If Charlie goes and Ben does not, then Diane will go but Alice will not.

Let us define A to be the proposition "Alice will go to the party", B to be "Ben will go to the party," etc. Then statements 1 through 3 above may be translated as follows:

	Conditional		Equation	
1.	$A \longrightarrow$	$B'C$	$AB + AC'$	$= 0$
2.	$BD \longrightarrow$	$A'C + AC'$	$BD(A'C' + AC)$	$= 0$
3.	$B'C \longrightarrow$	$A'D$	$AB'C + B'CD'$	$= 0$

The given data are therefore equivalent to the propositional equation $f = 0$, where f is given by

$$f = A(B + C') + BD(A'C' + AC) + B'C(A + D') . \qquad (5.25)$$

The Blake canonical form for f, i.e.,

$$BCF(f) = BC'D + B'CD' + A , \qquad (5.26)$$

is found from (5.25) by multiplying out to obtain an SOP formula, applying consensus repeatedly, and deleting absorbed terms. The prime clauses for the Halloween party are therefore the following:

$BD \longrightarrow C$	"If Bill and Diane go to the party,	
	then Charlie will go."	
$C \longrightarrow B + D$	"If Charlie goes to the party,	
	then either Bill or Diane will go."	
$A \longrightarrow 0$	"Alice will not go to the party."	

We show in Section 5.5 that clauses derived in this way constitute a complete and simplified representation of all conclusions that may be inferred from the given premises.

Example 5.4.2 The RS flip-flop is defined by the equations

$$Y = S + yR' \quad \text{(characteristic equation)}$$
$$0 = RS \quad \text{(input constraint)}$$

where R (Reset) and S (Set) are input-excitation signals and y and Y are the present state and next states, respectively. The foregoing equations are equivalent to the single equation $f = 0$, where $BCF(f)$ is given as follows:

$$BCF(f) = RS + RY + SY' + R'yY' + S'y'Y .$$

The associated prime clauses, viz.,

1.	RS	$\longrightarrow$	0
2.	RY	$\longrightarrow$	0
3.	S	$\longrightarrow$	Y
4.	y	$\longrightarrow$	$R + Y$
5.	Y	$\longrightarrow$	$S + y$

may be interpreted as follows:

1. "Set and Reset cannot be high simultaneously."

2. "Reset and the next state cannot be high simultaneously."

3. "If Set is high, the next state will be high."

4. "If the present state is high, then Reset is high or the next state will be high."

5. "If the next state will be high, then Set is high or the present state is high."

These statements, though doubtless not as intuitive to a designer as are the characteristic equation and input constraint, provide a standardized and complete specification for the RS flip-flop. The specification is complete in that any clause that can be inferred from the given premises is a superclause of one of the prime clauses. Thus all simplified deduced clauses are present among the prime clauses; clause 1 for example may be deduced from clauses 2 and 3.

5.5   Producing and Verifying Consequents

Let us consider two collections of logical data, one reducible to the equation $f = 0$ and the other reducible to the equation $g = 0$. We assume as before that the functions f and g are expressed as SOP formulas. As we have seen, the equation $g = 0$ is a consequent of $f = 0$ if and only if each term of g is included in some term of $BCF(f)$. Given $BCF(f)$, therefore, the task of *verifying* consequents of f becomes a matter of term-by-term comparison. The task of *producing* consequents may similarly be performed, as we now demonstrate, on a termwise basis.

5.5.1   Producing Consequents

To illustrate the process of producing consequents systematically, let us return to Example 5.0.1, concerning Alfred's collegiate endeavors. The premises reduce to the equation $f = 0$, where

$$BCF(f) = E'S' + G' . \tag{5.27}$$

An equation $g = 0$ is therefore an Alfred-consequent if and only if each term of g is included in either $E'S'$ or G'; we tabulate the possible terms of g, in terms of propositions E, G, and S, in Table 5.1.

Every function g forming a consequent $g = 0$ of the Alfred- premises (and nothing else) is assembled as the disjunction of a subset (possibly empty) of the eleven distinct terms enumerated in Table 5.1 (the term $E'G'S'$ appears twice). Some of the consequents thus assembled are the following:

$$
\begin{aligned}
E'S' + E'G' + G'S &= 0 \\
EG'S &= 0 \\
E'GS' + EG'S &= 0 \\
0 &= 0 .
\end{aligned}
$$

Terms included in $E'S'$	Terms included in G'
E'S'	G'
E'G'S'	E'G'
E'GS'	EG'
	G'S'
	G'S
	E'G'S'
	E'G'S
	EG'S'
	EG'S

Table 5.1: Terms included in $E'S' + G'$.

There are $2^{11} = 2048$ SOP g-formulas, distinct to within congruence, in the three letters E, G, and S that may be assembled in this way. The function f, however, covers 5 minterms on E, G, and S; thus there are only $2^5 = 32$ distinct g-functions included in f. Although there is redundancy from a functional standpoint, each of the 2048 g-formulas represents a distinct set of clauses deducible from the premises. The third of the four consequents above, for example, corresponds to the following set of clauses:

$$G \longrightarrow E + S$$ "If Alfred gets good grades
then Alfred enjoys college
or Alfred studies."

$$ES \longrightarrow G$$ "If Alfred enjoys college
and studies
then Alfred gets good grades."

5.5.2 Verifying Consequents

Let us decide whether the following proposition is a consequent of the Halloween-party premises of Example 5.4.1:

"If Alice and Ben both go to the party, or if neither of them goes, then Diane will go or Charlie will not go."

As a symbolic conditional:

$$A'B' + AB \longrightarrow C' + D . \tag{5.28}$$

As an equation:

$$A'B'CD' + ABCD' = 0 . \tag{5.29}$$

To verify that (5.29) is a valid consequent of the Halloween-party premises, we recall the Blake canonical form (5.26):

$$BCF(f) = BC'D + B'CD' + A .$$

The term $A'B'CD'$ of (5.29) is included in the term $B'CD'$ of $BCF(f)$; likewise, the term ABCD' of (5.29) is included in the term A of $BCF(f)$. Hence, $g \ll BCF(f)$, and thus $g \leq f$. We conclude therefore that the proposed consequent is valid.

5.5.3 Comparison of Clauses

The procedure we have just employed is to compare terms of g with terms of $BCF(f)$. We may also proceed by expressing a proposed consequent as a system of clauses, each of which we compare with the prime clauses. Let us suppose that terms p and q correspond, respectively, to clauses P and Q. Then p is included in q if and only if P is a superclause of Q, i.e., if and only if each letter appearing in the clause Q also appears, on the same side, in the clause P. The relevant clauses for the Halloween party are listed below.

Prime Clauses	Clauses of Proposed Consequent
$BD \longrightarrow C$	$C \longrightarrow A + B + D$
$C \longrightarrow B + D$	$ABC \longrightarrow D$
$A \longrightarrow 0$	

We observe that $C \longrightarrow A + B + D$ is a superclause of the prime clause $C \longrightarrow B + D$ and that $ABC \longrightarrow D$ is a superclause of the prime clause $A \longrightarrow 0$, verifying that the system of clauses shown on the right above is a valid consequent of the system of prime clauses shown on the left.

5.6 Class-Logic

Our examples thus far have been propositional. Let us now consider a problem in class-logic.

Example 5.6.1 On p. 112 of his *Symbolic Logic* [34], Lewis Carroll asks his readers to find conclusions deducible from the following premises:

(1) Babies are illogical.
(2) Nobody is despised who can manage a crocodile.
(3) Illogical persons are despised.

An appropriate universe is the set of human beings, among whose classes are the following: B = babies; M = able to manage a crocodile; D = despised; and L = logical. We write the premises as inclusions and as equations:

$$\begin{array}{ccc}
\multicolumn{3}{c}{\text{Inclusions}} \\
(1) & B \subseteq L' \\
(2) & M \subseteq D' \\
(3) & L' \subseteq D
\end{array} \qquad \begin{array}{ccc}
\multicolumn{3}{c}{\text{Equations}} \\
BL & = & 0 \\
MD & = & 0 \\
L'D' & = & 0 \,.
\end{array}$$

(We have used the set-notation $\subseteq$ for inclusion; the generic notation $\leq$ for inclusion in a Boolean algebra may also be used.)

The premises are equivalent to the single Boolean equation $f = 0$, where $f = BL + MD + L'D'$. Converting f to Blake canonical form, we obtain

$$BCF(f) = BL + MD + L'D' + BD' + ML' + BM \,.$$

The prime consequents, in clausal form, are the following:

$$\begin{array}{llcl}
(a) & BL & \subseteq & 0 \\
(b) & MD & \subseteq & 0 \\
(c) & 1 & \subseteq & L + D \\
(d) & B & \subseteq & D \\
(e) & M & \subseteq & L \\
(f) & BM & \subseteq & 0 \,.
\end{array}$$

Prime consequents (a), (b), and (c) are the clausal forms, respectively, of premises (1), (2), and (3). Prime consequents (d), (e), and (f) may be given the following interpretations:

(d) Babies are despised.
(e) Anybody who can manage a crocodile is logical.
(f) No baby can manage a crocodile.

All of the premises survive as prime consequents because Carroll's example has a specialized logical form, called "sorites," which may be resolved into a chain of simple inclusions. The inclusion-chain for this example is

$$B \subseteq L' \subseteq D \subseteq M' \,.$$

5.7   Selective Deduction

An important class of logical problems involves selective deduction from given hypotheses. The following example is a modification of one given by Ledley [113].

Example 5.7.1 Enzyme biochemistry has two characteristic features. First, it is usually difficult to isolate an enzyme in pure form, and thus the chemist must deal with imprecise and indirect knowledge of the enzyme content of the experimental ingredients. Second, usually more than one chemical reaction takes place at once, and even these are observed indirectly. Suppose a chemist is studying enzymes A, B, and C in relation to reactions X, Y, and Z. He has completed the following experiments:

1. In the first experiment, a solution containing neither A, B, nor C produced reaction Y but neither X nor Z.

2. In the second experiment, the solution contained A and either B or C or both (the chemist could not be sure); the reaction was neither Y nor was it X and Z together.

3. In the third experiment, the solution had B but not A, or did not have B but had C. Reactions X and Y occurred, or reaction X did not occur but Z did.

4. In the fourth experiment, the chemist obtained a solution from a source that had C, together with A or B or both, or else had neither A nor C. Either reaction X did not take place, or both Y and Z did.

5. In the fifth experiment, a solution containing A but not B either failed to produce reaction X or failed to produce reaction Z.

Having made the foregoing observations, the chemist seeks answers, in the simplest possible form, to the following questions:

(a) What is known concerning the reactions X, Y, and Z, independent of any knowledge of enzyme content?

(b) What is known concerning the enzymes A, B, and C, given each of the following reactions?

> i) X occurred;
> ii) X did not occur;
> iii) Z occurred;
> iv) Z did not occur.

Our approach will be to reduce the information provided by experiments 1 through 5 to a single equation $f = 0$, and then to express the function f in Blake canonical form. This form enables the chemist to eliminate conveniently the variables not of current interest.

The experimental information is expressed by the following system of conditionals:

$$
\begin{array}{lll}
(1) & A'B'C' & \longrightarrow & X'YZ' \\
(2) & A(B+C) & \longrightarrow & Y'(X'+Z') \\
(3) & A'B+B'C & \longrightarrow & XY+X'Z \\
(4) & C(A+B)+A'C' & \longrightarrow & X'+YZ \\
(5) & AB' & \longrightarrow & X'+Z'
\end{array}
$$

This system is equivalent to the single equation $f = 0$, where f is expressed by

$$
\begin{aligned}
f \;=\; & A'B'C'(X+Y'+Z)+ \\
& +A(B+C)(Y+XZ)+ \\
& +(A'B+B'C)(XY'+X'Z')+ \\
& +(AC+BC+A'C')X(Y'+Z')+ \\
& +AB'XZ.
\end{aligned}
$$

In Blake canonical form:

$$
\begin{aligned}
BCF(f) \;=\; & ACX+AXZ+ACY+AB'CZ'+ABY+A'XY'+CXY'+ \\
& +XY'Z+A'Y'Z'+B'CY'Z'+A'CX'Z'+CX'YZ'+ \\
& +B'CX'Z'+A'B'C'X+B'C'XZ+A'B'C'Y'+A'B'C'Z+ \\
& +A'BZ'+BCXZ'+BYZ'+A'C'XZ'.
\end{aligned}
$$

To answer question (a) posed by the chemist, we eliminate A, B, and C from the equation $f = 0$. This may be done simply by deleting from the equation $BCF(f) = 0$ every term involving A, B, or C (cf. Theorem 4.8.2). The result,

$$
XY'Z = 0,
$$

takes the clausal form

$$XZ \longrightarrow Y .$$

Thus the following is known, independent of any knowledge of enzyme content: if reactions X and Z occur together, then reaction Y occurs also. To answer the first and second parts of question (b), we eliminate Y and Z from $BCF(f) = 0$; to answer the third and fourth parts of (b), we eliminate X and Y. The resultants of elimination are:

Eliminating Y and Z: $\qquad A'B'C'X + ACX \;\; = \;\; 0$
Eliminating X and Y: $\quad AB'CZ' + A'B'C'Z + A'BZ' \;\; = \;\; 0$

The foregoing equations enable us to answer the chemist's question (b), in clausal form, as follows:

$$\begin{array}{rlcl}
\text{i)} & AC & \longrightarrow & 0 \\
& 1 & \longrightarrow & A + B + C \\
\text{ii)} & \text{NO INFORMATION} \\
\text{iii)} & 1 & \longrightarrow & A + B + C \\
\text{iv)} & AC & \longrightarrow & B \\
& B & \longrightarrow & A
\end{array}$$

5.8   Functional Relations

Suppose we wish to look for relations among a collection $f_1, f_2, \ldots, f_m$: $B^n \longrightarrow B$ of Boolean functions. Let the functions in such a collection be represented, respectively, by formulas $F_1, F_2, \ldots, F_m$ on the argument-vector $X = (x_1, \ldots, x_n)$. The system

$$\begin{array}{rcl}
A_1 & = & F_1(X) \\
A_2 & = & F_2(X) \\
& \vdots & \\
A_m & = & F_m(X)
\end{array} \qquad (5.30)$$

associates the symbols in the vector $A = (A_1, \ldots, A_m)$ with the corresponding formulas. All of the relations implied among the original functions are therefore encoded economically in those prime consequents of (5.30) which do not involve any of the X-arguments; let us call these the *A-consequents* of (5.30). Such consequents are equations whose right-hand sides are zero; let us call their left-hand sides the *A-consequent terms*.

The label-and-eliminate procedure.   The A-consequent terms of (5.30) are generated by reducing (5.30) to a single equivalent equation of the form $g(A, X) = 0$, expanding $g(A, X)$ into Blake canonical form, and selecting those terms not involving X-variables. Computational efficiency is improved if the X-variables are eliminated prior to the Blake-expansion. In detail:

Step 1. Reduce (5.30) to the single equivalent equation

$$g(A, X) = 0 .\qquad(5.31)$$

Step 2. Eliminate X from (5.31), yielding the resultant

$$ECON(g(A, X), X) = 0 .\qquad(5.32)$$

Step 3. Express the left side of (5.32) in Blake canonical form, i.e.,

$$BCF(ECON(g(A, X), X)) = 0 .\qquad(5.33)$$

Example 5.8.1 Consider the Boolean functions labelled by the system

$$
\begin{aligned}
A_1 &= x_1 + x_2 \\
A_2 &= x_1 \\
A_3 &= x_1 x_2 \\
A_4 &= x_2'
\end{aligned}
\qquad(5.34)
$$

The output of a program to derive the A-consequent terms from (5.34) is listed below:

```
A1 A2'A4
A2'A3
A2 A3'A4'
A1'A2
A1'A3
A3 A4
A1'A4'
```

The corresponding clauses, i.e., the prime clauses of (5.34), are as follows:

```
A1 A4  ---> A2
A3     ---> A2
A2     ---> A3 + A4
A2     ---> A1
A3     ---> A1
A3 A4  ---> 0
1      ---> A1 + A4
```

These seven clauses, corresponding to the terms on the left side of (5.33), constitute a simplified and complete representation of the relations holding among the original functions. Typically only relations of specialized form are sought; three such specializations are discussed in the following sections.

5.9   Dependent Sets of Functions

Questions concerning the dependence of collections of sets, propositions, or Boolean functions have been widely investigated; see Marczewski [128] for citations to early work. A set of k propositional or switching functions is customarily said to be *independent* in case all 2^k combinations of values are possible. This interpretation of independence has been applied to the design of switching circuits by Muller [141], Kjellberg [102], and Ledley [118]. Ledley has applied it also to problems in enzyme biochemistry [115]. Kuntzmann [110] and Small [189] have employed this interpretation to arrive at results concerning the decomposition of switching functions.

The foregoing interpretation is applicable to two-valued sets such as those comprising axioms, propositions, or switching functions, but is inadequate for sets of Boolean functions on an arbitrary Boolean algebra. Let B be such a Boolean algebra, let $T = \{f_1, f_2, \ldots, f_m\}$ be a set of Boolean functions mapping B^n into B, and let $S = \{f_1, f_2, \ldots, f_k\}$ be a subset of T (the first k elements of T are selected, without loss of generality). As in Brown & Rudeanu [26, 29] (see also Marczewski [128]), we call the subset S *functionally dependent* provided there is a non-constant Boolean function $h: B^k \longrightarrow B$ for which the identity

$$h(f_1(X), \ldots, f_k(X)) = 0 \qquad (\forall X \in B^n) \qquad (5.35)$$

is fulfilled; otherwise S will be called *functionally independent*.

We consider two problems concerning the set T. The first is to establish whether a given subset of T is dependent. The second is to produce an economical representation for the family of all dependent subsets of T and the complementary family of all independent subsets.

Every subset of of an independent set is independent; in particular, the empty set is independent. Every superset of a dependent set is also dependent; if the family of dependent subsets is not empty, therefore, the entire set T is dependent. An independent set is *maximal* in case there is no independent set strictly including it; a dependent set is *minimal* in case there is no dependent set strictly included in it. A subset of T is dependent (independent), therefore, if and only if it includes a minimal dependent set (it is included in a maximal independent set). Let us denote by $IMAX$ the class of maximal independent subsets of T and by $DMIN$ the class of minimal dependent subsets of T.

It follows from our definition of functional dependence and from the discussion in Section 5.8 that a subset S of T is dependent if and only if all of the letters in an A-term derived from (5.30) belong to the set $\{A_1, A_2, \ldots, A_k\}$ of letters associated with S.

To derive the maximal independent and minimal dependent subsets of T, we construct a complement-free SOP formula W from (5.33) as follows. If $BCF(ECON(g(A,X),X))$ is null, then W is defined to be null; otherwise, the terms of W are formed in one-to-one correspondence with the terms of $BCF(ECON(g(A,X),X))$ by

(i)   deleting all constants (elements of B), and
(ii) replacing either A_i or A'_i by A_i $(i = 1, 2, \ldots, m)$.

Each term of the formula $BCF(ECON(g(A,X),X))$ contains at least one A-letter; hence step (i) cannot annihilate a term.

Let $w : \{0,1\}^n \longrightarrow \{0,1\}$ be the Boolean function represented by formula W; we call w the *dependency function* associated with the system (5.30). In the case of independence (i.e., internal stability) of the vertices of a graph, the complement w' of w is the Boolean function introduced by Maghout [126, 127] and Weissman [211].

The following result is proved in Brown & Rudeanu [29]:

Theorem 5.9.1 *Let* $T = \{f_1, f_2, \ldots, f_m\}$ *be a set of Boolean functions, let* $S = \{f_1, \ldots, f_k\}$ *be a subset of* T, *let* $S' = T - S$ *be the set-complement of* S *relative to* T *and let* w *be the function defined above. Then*

(a) *S is a minimal dependent subset of T if and only if $A_1 \cdots A_k$ is a term of $BCF(w)$.*

(b) *S' is a maximal independent subset of T if and only if $A'_1 \cdots A'_k$ is a term of $BCF(w')$.*

Example 5.9.1 Let us find the maximal independent and minimal dependent subsets of the set $T = \{f_1, f_2, f_3, f_4\}$ of Boolean functions specified by the system

$$
\begin{aligned}
A_1 &= bx + y \\
A_2 &= bx \\
A_3 &= x + b'z \\
A_4 &= y
\end{aligned}
$$

on the Boolean algebra $B = \{0, 1, b', b\}$.

We first produce the set of A-terms:

```
A1 A3'A4'
A1 A4'B'
A1 A2'A4'
A1'A4
A2'A3 B
A2 A3'
A2 B'
A1'A2
A1'A3 B
```

Hence

$$
\begin{aligned}
BCF(w) &= A_1 A_3 + A_2 + A_1 A_4 \\
BCF(w') &= A'_2 A'_3 A'_4 + A'_1 A'_2 .
\end{aligned}
$$

We conclude from Theorem 5.9.1 that

$$
\begin{aligned}
DMIN &= \{\{f_1, f_3\}, \{f_2\}, \{f_1, f_4\}\} \\
IMAX &= \{\{f_1\}, \{f_3, f_4\}\} .
\end{aligned}
$$

The family $DMIN$ generates 11 dependent subsets of $\{f_1, f_2, f_3, f_4\}$; the family $IMAX$ generates 5 independent subsets.

5.10   Sum-to-One Subsets

The tautology problem is discussed in Section 4.11. An associated problem arising in a number of applications is to find all minimal sum-to-one subsets of a set $T = \{t_1(X), t_2(X), \ldots, t_m(X)\}$ of terms (products). Such subsets correspond to the A-terms implied by (5.30) consisting entirely of complemented literals. In Example 5.8.1, therefore, there is only one minimal sum-to-one subset, corresponding to the A-term $A_1' A_4'$ (if $A_1' A_4' = 0$, then $A_1 + A_4 = 1$).

It is readily verified that the subset $\{f_1, f_4\}$ sums to one and no other subsets (save supersets of $\{f_1, f_4\}$) sum to one.

Although the process just discussed (searching for A-terms of (5.30) consisting entirely of complemented literals) generates minimal sum-to-one subsets, it is an unnecessarily complex way to do the job. A more direct approach is based on the auxiliary summation

$$A_1 t_1(X) + A_2 t_2(X) + \ldots + A_m t_m(X) .$$

The utility of this formula is based on the following result:

Lemma 5.10.1 *Let* $T = \{t_1(X), t_2(X), \ldots, t_m(X)\}$ *be a set of terms, where* $X = (x_1, \ldots, x_n)$, *and let* $A_1, A_2, \ldots, A_m$ *be Boolean variables. Let* $S = \{t_1(X), t_2(X), \ldots, t_k(X)\}$ *$(1 \le k \le m)$ be a subset of* T. *Then the conditions*

$$\text{(a)} \quad t_1(X) + t_2(X) + \ldots + t_k(X) = 1$$
$$\text{(b)} \quad A_1 A_2 \cdots A_k \le A_1 t_1(X) + \ldots + A_m t_m(X)$$

are equivalent for all $X \in \{0, 1\}^n$.

Proof.   Condition (b) is equivalent to the equation

$$A_1' + \ldots + A_k' + A_1 t_1(X) + \ldots + A_m t_m(X) = 1 ,$$

which may be expressed equivalently as

$$\begin{aligned} A_1' + \ldots + A_k' + t_1(X) + \ldots + t_k(X) + \\ + A_{k+1} t_{k+1}(X) + \ldots + A_m t_m(X) \quad &= \quad 1 . \end{aligned} \qquad (5.36)$$

Thus (a) implies (b). To show that (b) implies (a) for all $X \in \{0, 1\}^n$, let us suppose that there is a member, K, of $\{0, 1\}^n$ such that (a) is false, i.e., for which $t_1(K) + \ldots + t_k(K) = 0$. Then (5.36) becomes

$$A_1' + \ldots + A_k' + A_{k+1} t_{k+1}(K) + \ldots + A_m t_m(K) = 1 ,$$

which is not an identity; thus (b) is false. Hence (b) implies (a). $\square$

Constructing Minimal sum-to-one subsets. Lemma 5.10.1 shows that the problem of finding sum-to-one subsets of a set T of terms reduces to that of finding A-products included in a summation associated with T. Any superset of a sum-to-one subset is a sum-to-one subset; hence, the entire collection of sum-to-one subsets is generated conveniently by the minimal sum-to-one subsets. The task of finding such subsets is eased by the following theorem:

Theorem 5.10.1 *Let* $T = \{t_1(X), t_2(X), \ldots, t_m(X)\}$ *be a set of terms, where* $X = (x_1, \ldots, x_n)$, *and let* $A_1, A_2, \ldots, A_m$ *be Boolean variables. Let* $S = \{t_1(X), t_2(X), \ldots, t_k(X)\}$ $(1 \le k \le m)$ *be a subset of* T. *Then* S *is a minimal sum-to-one subset of* T *if and only if the product* $A_1 A_2 \cdots A_k$ *is a term of* $BCF(A_1 t_1(X) + \ldots + A_m t_m(X))$.

Proof.   Denote by $G(A, X)$ the function $A_1 t_1(X) + \ldots + A_m t_m(X)$. The following statements are equivalent:

    (i) S is a sum-to-one subset of T.
    (ii) $A_1 A_2 \cdots A_k \le G(A, X)$.
    (iii) $A_1 A_2 \cdots A_k \le p$, where p is a term of $BCF(G(A, X))$.

The equivalence between (i) and (ii) follows from Lemma 5.10.1; that between (ii) and (iii) from Lemma A.3.2. Thus S is a sum-to-one subset of T if and only if a subproduct of $A_1 A_2 \cdots A_k$ is a term of $BCF(G(A, X))$. The set S is therefore a minimal sum-to-one subset of T if and only if $A_1 A_2 \cdots A_k$ is a term of $BCF(A_1 t_1(X) + \ldots + A_m t_m(X))$. $\square$

Example 5.10.1 Consider the collection of terms named in the system

$$
\begin{aligned}
t_1 &= x \\
t_2 &= x'y \\
t_3 &= y'z' \\
t_4 &= z \\
t_5 &= x'y' \\
t_6 &= yz'
\end{aligned}
$$

We derive from this system the formula

$$A_1 x + A_2 x'y + A_3 y'z' + A_4 z + A_5 x'y' + A_6 yz'$$

and generate the associated a-terms as follows:

A1 A2 A3 A4
A1 A2 A5
A1 A4 A5 A6
A3 A4 A6

The minimal sum-to-one subsets, corresponding to the foregoing A-terms, are $\{f_1, f_2, f_3, f_4\}$, $\{f_1, f_2, f_5\}$, $\{f_1, f_4, f_5, f_6\}$, and $\{f_3, f_4, f_6\}$.

5.11   Irredundant Formulas

The problem of minimizing the complexity of Boolean formulas is important in technology, and has received extensive attention in the literature. We show in this section that minimal sum-of-products (SOP) formulas for a Boolean function may be generated via syllogistic reasoning.

Let us review, from Section 4.11, the major points concerning the problem of finding a simplified SOP formula for a Boolean function f. Such a formula is necessarily a subformula of $BCF(f)$. An *irredundant* formula for f is a subformula of $BCF(f)$ that (a) represents f and (b) ceases to represent f if any of its terms is deleted. The search for simplified SOP formulas therefore need only be over the irredundant formulas.

Quine [161] presented a tabular method for finding all of the irredundant formulas for a given function. Algebraic alternatives have been suggested by Samson & Mueller [175], Petrick [154], Ghazala [70], Mott [140], Cutler & Muroga [43] and others.

Incompletely-specified functions.   Associated with the foregoing problem is the more general problem of finding simplified SOP formulas for an *incompletely specified* Boolean function f. Such a function is defined by an interval, i.e.,

$$g(X) \leq f \leq h(X) \tag{5.37}$$

(*cf.* Section 2.4) in which g and h are given Boolean functions. Each function in the interval (5.37) is represented by a set of irredundant SOP formulas; call it the *I-set* for that function. We define a formula to be an irredundant formula for the incompletely-specified function f provided (a) it belongs to one of the I-sets associated with the interval (5.37) and (b) none of its proper subformulas belongs to such an I-set.

It is a well-known result in switching theory that the irredundant formulas for the incompletely-specified function (5.37) are the minimal subformulas of $BCF(h)$ that cover g; therefore an SOP formula S is an irredundant formula for (5.37) if and only if

(a) S is a subformula of $BCF(h)$,
(b) $g \leq S$, and
(c)   no proper subformula of S has property (b).

All of the formulas satisfying the foregoing conditions are readily found by syllogistic reasoning, using a variation of the label-and-eliminate technique described in Section 5.8.

Theorem 5.11.1 *Let g and h be the Boolean functions specified by interval (5.37), let $p_1, \ldots, p_m$ be the prime implicants of h, let $\phi = 0$ be equivalent to the system*

$$
\begin{aligned}
u &= g(X) \\
p_1(X) &= A_1 \\
&\ \vdots \\
p_m(X) &= A_m \ ,
\end{aligned}
$$

(5.38)

and let $A_{\alpha_1}, \ldots, A_{\alpha_k}$ be symbols in the set $\{A_1, \ldots, A_m\}$. Then the SOP formula

$$ p_{\alpha_1} + \ldots + p_{\alpha_k} $$

(5.39)

is an irredundant formula for (5.37) if and only if

$$ A'_{\alpha_1} \cdots A'_{\alpha_k} u $$

(5.40)

is a prime implicant of ϕ.

Proof. Formula (5.40) is a prime implicant of ϕ if and only if the relation $u \leq A_{\alpha_1} + \ldots + A_{\alpha_k}$ is a prime consequent of system (5.38), i.e., if and only if the function g satisfies the conditions

(i) $g \ \leq \ p_{\alpha_1} + \ldots + p_{\alpha_k}$
(ii) $g \ \not\leq \ $ any proper subformula of $p_{\alpha_1} + \ldots + p_{\alpha_k}$,

where $p_{\alpha_1} + \ldots + p_{\alpha_k}$ is a subformula of $BCF(f)$. $\square$

The process of deduction is simplified if the X-variables are eliminated from (5.38) before the prime implicants are sought, i.e., if we follow the label-and-eliminate procedure of Section 5.8.

Example 5.11.1 Let an incompletely-specified function f be defined by the interval (5.37), where g and h are given by the formulas

$$g(x, y, z) = x'yz + xy'$$
$$h(x, y, z) = x'z + xy' + yz' .$$

Thus (5.38) takes the form

$$
\begin{aligned}
u &= x'yz + xy' \\
x'z &= A_1 \\
xy' &= A_2 \\
yz' &= A_3 \\
y'z &= A_4 \\
x'y &= A_5 \\
xz' &= A_6 .
\end{aligned}
\tag{5.41}
$$

System (5.41) is equivalent to an equation $\phi(A_1, \ldots, A_6, u, x, y, z) = 0$ for which the prime implicants of $ECON(\phi, \{x, y, z\})$ are

A1 A4'U'	A2'A6 U	A5 A6
A1'A4 U'	A3 U	A1 A2
A2 U'	A1 A5'U	A2 A3
A3'A5 U'	A1'A5 U	A2 A5
A3'A6 U'	A1'A2'U	A2 A4'A6'
A1 A5 U'	A1 A6	A3 A4
A1'A4'A6'U	A1 A3	A4 A6
A4'A5'A6'U	A2'A3'A6	A4 A5
A1 A4 U	A1'A3'A5	A1'A2'A4
A2'A4 U	A3 A5'A6'	A1 A4'A5'
A2'A5'U		

The prime implicants corresponding to irredundant formulas are those having the form $A'_{\alpha_1} \cdots A'_{\alpha_k} u$, viz., $A'_1 A'_4 A'_6 u$, $A'_4 A'_5 A'_6 u$, $A'_2 A'_5 u$, and $A'_1 A'_2 u$. Thus the irredundant formulas in the given interval are

$$
\begin{aligned}
&x'z + y'z + xz' \\
&y'z + x'y + xz' \\
&x'y' + x'y \\
&x'z + x'y' .
\end{aligned}
$$

Of the 31 prime implicants in the foregoing example, only four have the form corresponding to an irredundant formula. A modification of system (5.38) leads to more economical results:

Corollary 5.11.1 *Let g, h and $p_1, \ldots, p_m$ be as defined in Theorem 5.11.1 and let $\hat{\phi} = 0$ be the equation to which the system*

$$
\begin{aligned}
u &\leq g(X) \\
p_1(X) &\leq A_1 \\
&\vdots \\
p_m(X) &\leq A_m
\end{aligned}
\tag{5.42}
$$

reduces. Then the prime implicants of $ECON(\hat{\phi}, X)$ are in one-to-one correspondence, in the manner described in Theorem 5.11.1, to the irredundant formulas in the interval (5.37).

Proof. The proof is left as an exercise.

Example 5.11.2 For the incomplete function f specified in Example 5.11.1, system (5.42) takes the form

$$
\begin{aligned}
u &\leq x'yz + xy' \\
x'z &\leq A_1 \\
xy' &\leq A_2 \\
yz' &\leq A_3 \\
y'z &\leq A_4 \\
x'y &\leq A_5 \\
xz' &\leq A_6 \,,
\end{aligned}
\tag{5.43}
$$

which is equivalent to $\hat{\phi} = 0$, where

$$
\begin{aligned}
\hat{\phi} = \;& ux'y' + ux'z' + uxy + \\
& + A_1'x'z + A_2'xy' + A_3'yz' + A_4'y'z + A_5'x'y + A_6'xz' \,.
\end{aligned}
$$

Hence, in Blake canonical form,

$$
ECON(\hat{\phi}, \{x, y, z\}) = A_1'A_2'u + A_1'A_4'A_6'u + A_2'A_5'u + A_4'A_5'A_6'u \,.
$$

The prime implicants of $ECON(\hat{\phi}, \{x, y, z\})$ are thus precisely those found in Example 5.11.1 to correspond to irredundant formulas.

Completely-specified functions. An ordinary (completely-specified) function may be regarded as a one-element interval of the form (5.37), i.e., one for which $g(X) = h(X)$. The procedure for generating the irredundant formulas for an ordinary function is therefore identical to that for an incompletely-specified function.

Exercises

1. Five workers—V, W, X, Y, and Z—are available to perform a certain task. In choosing a hiring-list, the following conditions must be satisfied:

   (a) Either X and Y are both hired, or neither is hired.

   (b) At least one of V, X, or Z must be hired.

   (c) If V is hired, then X or Y (or both) must be hired.

   (d) If X and Y are hired, then Z must be hired.

   (e) If Z is not hired, then W must be hired.

   Express the prime consequents in clausal form.

2. The state of a mechanism under test is shown by 5 indicators, labelled A, B, C, D, and E. After watching the indicators for a long time, an observer characterizes the mechanism as follows:

   (a) If A or D is on (but not both), then C is on.

   (b) Looking just at C, D, and E, the number of on-indicators is always odd.

   (c) If E is off, then A and D are both off.

   (d) If B and C are both on, then E is on.

   (e) At least one of the following conditions always exists:

      i. A on.

      ii. C off.

      iii. D on.

   Express the prime consequents in clausal form.

3. Test the validity of the following argument.

PREMISES   CONCLUSION

$$PQ \longrightarrow (R+S)(R'+S') \qquad P \longrightarrow Q'$$
$$S \longrightarrow (QR+Q'R')+P'$$
$$QR \longrightarrow S$$

4. Let P, Q, R, and S be elements of a Boolean algebra and suppose we are given the following premises:

- $Q + R = 1$
- $P' + Q = P'R' + PQ$
- If $PQ' = 1$, then $R' + S' = 1$.
- $QS = 0$
- If $P = Q$, then $R = 1$.

(1) List the prime consequents.

(2) Test the validity of each of the following proposed consequents of the given premises.

  (a) If $Q = 1$, then $P = R'$.
  (b) $P + Q = R$
  (c) If $P = R$, then $P + Q = 0$.
  (d) $P'S'(QR + Q'R') = 0$.

5. (Keynes [101], Part IV) At a certain examination,

  (a) all the candidates who were entered for Latin were also entered for either French, German, or Spanish, but not for more than one of these languages;

  (b) all the candidates who were not entered for German were entered for at least two of the other languages; and

  (c) no candidate who was entered for both French and Spanish was entered for German, but all candidates who were entered for neither French nor Spanish were entered for Latin.

Show that each candidate was entered for exactly two of the four languages.

6. The following problem, given in Chapter IX of Boole's *Laws of Thought*, was used as an example by Schröder [178], Venn [210], Peirce [151], Ladd [111], and other nineteenth-century logicians; Schröder called it the "touchstone" for his work. It was also used by Blake [10] in the first published example of the generation of prime implicants.

> Suppose that an analysis of properties a, b, c, d, and e of a particular class of substances leads to the following statements:
>
> (1) Whenever properties a and c are missing, then property e is found, together with one of the properties b and d, but not both.
>
> (2) Whenever the properties a and d are found while e is missing, then both b and c will either both be found or both be missing.
>
> (3) Whenever property a is found in conjunction with either b or e, or both of them, then c or d will also be found, but not both of them. Conversely, whenever c or d (but not both) is found, then a will be found in conjunction with either b or e or both of them.

(a) Reduce all of the foregoing data to a single Boolean equation of the form $f(a, b, c, d, e) = 0$.

(b) Construct BCF(f).

(c) What independent relations among b, c, and d may be inferred? What among these three may be inferred if $a = 1$?

(d) What independent relations among a, c, and d may be inferred? What with the further hypothesis $b = 1$?

(e) Which of the following is a valid consequent of the premises (1), (2), (3)?

    i. $ad \longrightarrow b'e' + c'e$

    ii. $a'c' \longrightarrow be$

    iii. $a + bc \longrightarrow ce$

    iv. $abc \longrightarrow 0$

7. Suppose that our knowledge of aardvarks (class A), creatures who kiss babies (K), courteous creatures (C), and politicians (P) is expressed by the following statements:

   (a) Politicians who do not kiss babies are courteous.
   (b) Courteous aardvarks are not politicians.
   (c) Aardvarks who kiss babies are courteous politicians.
   (d) All politicians either kiss babies, or are aardvarks, or both.

   State all of the prime consequents, in clausal form, of the foregoing data.

8. (Venn [210], Chapter XIII) There is a certain class of things from which A picks out the X that is Z and the Y that is not Z, and B picks out from the remainder the Z which is Y and the X that is not Y. It is then found that what is left exactly comprises Z which is not X.

   (a) State the implied constraint (if any) relating X, Y, and Z.
   (b) Assuming the constraint to be satisfied, what can be determined about the original class?

9. The RST flip-flop is defined by the equations

$$Y = S + yR'T' + y'T \quad \text{(characteristic equation)}$$
$$0 = RS + RT + ST \quad \text{(input constraint)}$$

   where y and Y are the present state and next states, respectively, of the flip-flop. List the prime consequents in clausal form. Each such consequent represents a fundamental property of the RST flip-flop.

10. (Corollary 5.11.1) Let g, h and $p_1, \ldots, p_m$ be as defined in Theorem 5.11.1 and let $\hat{\phi} = 0$ be the equation to which the system

$$u \leq g(X)$$
$$p_1(X) \leq A_1$$
$$\vdots$$
$$p_m(X) \leq A_m$$

   reduces. Show that the prime implicants of $ECON(\hat{\phi}, X)$ are in one-to-one correspondence, in the manner described in Theorem 5.11.1, to the irredundant formulas in the interval (5.37).

Chapter 6

Solution of Boolean Equations

Many problems in the application of Boolean algebra may be reduced to that of solving a Boolean equation of the form

$$f(X) = 0 , \tag{6.1}$$

over a Boolean algebra B. The specifications for a digital circuit, for example, typically take the form of Boolean equations relating a collection $X = (x_1, x_2, \ldots, x_n)$ of output variables to a collection $I = (i_1, i_2, \ldots, i_r)$ of input-symbols. These specifications may be reduced (by Theorem 4.3.1) to a single equivalent equation of the form (6.1) over the free Boolean algebra (cf. Section 2.13.1) generated by the input-symbols. The designer's task is to construct a system

$$
\begin{aligned}
x_1 &= g_1 \\
&\vdots \\
x_n &= g_n
\end{aligned}
\tag{6.2}
$$

of Boolean equations in which $g_1, \ldots, g_n$ are formulas in $FB(i_1, \ldots, i_r)$ that (a) specify circuit-structure and (b) accord with the original specifications. To meet the latter condition, system (6.2) should imply equation (6.1), i.e., it should be an antecedent of (6.1) (cf. Section 4.2). An antecedent system of the form (6.2) is a *functional antecedent* , i.e., a *solution* , of (6.1). The latter term is more common and will be used henceforth.

Formal procedures for producing solutions of (6.1) were developed by Boole himself as a way to treat problems of logical inference, and Boolean equations have been studied extensively since Boole's initial work. See Rudeanu [172] for a comprehensive modern treatment of Boolean equations and a bibliography of nearly 400 sources; among the classical sources are Schröder's three-volume text [178] and Couturat's brief and lucid monograph [41].

Boolean equation-solving as an approach to the design of relay-networks was discussed by Nakasima [145, 146] as early as 1936, and later by Ashenhurst [4], Semon [181], and Ledley [113]. Ledley considered applications to the design of gate-networks (adders and squaring circuits); he also discussed applications in medical diagnosis, enzyme biochemistry, and agricultural experiments. Applications in the design of digital computers were first suggested in the texts by Phister [155] and Ledley [118]; other work has been done by Cerny and Marin [36] and by Svoboda and White [193].

Among the many fields to which Boolean equation-solving has been applied systematically are biology, grammars, graph theory, chemistry, law, medicine and spectroscopy. Klir and Marin [103] have noted of Boolean equations that "their importance for switching theory reminds one of the application of differential equations in electric circuit theory." Applications in logical design include functional decomposition, fault-diagnosis, binary codes, a variety of approaches to combinational synthesis, flip-flop design and excitation, hazard-free synthesis, information-lossless machines, and the design of sequential circuits (the latter application has given rise to a specialized field of investigation, viz., sequential Boolean equations). An extensive survey of applications is given in Rudeanu [172].

6.1　Particular Solutions and Consistency

A *particular solution* (or, simply, a *solution*) of (6.1) is an element $A = (a_1, \ldots, a_n)$ of B^n such that $f(A) = 0$ is an identity. A Boolean equation is *consistent* provided it has at least one solution.

The one-variable Boolean equation $f(x) = 0$ is an important special case, whose consistency we now study. We recall from Sections 4.7 and 4.8 that the resultant of elimination of x from the equation $f(x) = 0$ is the equation $ECON(f, \{x\}) = 0$.

Lemma 6.1.1 *The Boolean equation $f(x) = 0$ is consistent if and only if the condition*

$$ECON(f, \{x\}) = 0 \tag{6.3}$$

is satisfied.

Proof. Suppose $a \in B$ is a solution of $f(x) = 0$, i.e., suppose that $f(a) = 0$. Then $a'f(0) + af(1) = 0$, by Boole's expansion theorem, whence $a'f(0) + af(1) + f(0)f(1) = 0$ by consensus. Thus $f(0)f(1) = 0$, i.e., $ECON(f, \{x\}) = 0$. Suppose on the other hand that $ECON(f, \{x\}) = 0$. Then the element $f(0)$ is a solution of $f(x) = 0$, for

$$f(f(0)) = (f(0))'f(0) + f(0)f(1) = 0 + ECON(f(x), \{x\}) = 0 + 0 = 0 .$$

Thus $f(x) = 0$ is consistent, proving the theorem. $\Box$

We call the equation $ECON(f, \{x\}) = 0$ the *consistency condition* for $f(x) = 0$. The consistency condition for an n-variable Boolean equation is a direct extension, as we now show, of that for a one-variable equation.

Theorem 6.1.1 *The n-variable Boolean equation (6.1) is consistent if and only if the condition*

$$ECON(f, \{x_1, \ldots, x_n\}) = 0 \tag{6.4}$$

is satisfied.

Proof. Suppose (6.1) is consistent, i.e., $f(A) = 0$ for some n-tuple A in B^n. Then (6.4) follows by Theorem 4.8.4. Conversely, suppose that condition (6.4) is satisfied. We show by induction that (6.1) is consistent for all $n \geq 1$. If $n = 1$, then (6.1) is consistent by Lemma 6.1.1. Suppose (6.1) to be consistent if $n = k$ ($k > 1$) and consider the $(k+1)$-variable equation $f(x_1, \ldots, x_k, x_{k+1}) = 0$. Condition (6.4) then takes the form

$$ECON(f, R \cup \{x_{k+1}\}) = 0 , \tag{6.5}$$

where $R = \{x_1, \ldots, x_k\}$. Let us define $g: B^k \longrightarrow B$ by

$$g(x_1, \ldots, x_k) = f(x_1, \ldots, x_k, 0)f(x_1, \ldots, x_k, 1) .$$

Condition (6.5) implies that $ECON(ECON(f, \{x_{k+1}\}), R) = 0$, which implies by the induction hypothesis that $ECON(f, \{x_{k+1}\}) = 0$ is consistent, i.e., that $g(a_1, \ldots, a_k) = 0$ for some k-tuple $(a_1, \ldots, a_k)$ in B^k. Therefore $f(a_1, \ldots, a_k, 0)f(a_1, \ldots, a_k, 1) = 0$, whence by Lemma 6.1.1 the equation $f(a_1, \ldots, a_k, x_{k+1}) = 0$ has a solution, call it a_{k+1}, in B. Thus the equation $f(x_1, \ldots, x_k, x_{k+1}) = 0$ is consistent. $\Box$

6.2   General Solutions

A *general solution* of a Boolean equation is a representation of the set, S, of its particular solutions. Although S may be represented by an explicit list, such a representation may obscure regularities in the form of the solutions. The number of solutions, moreover, may be so large that enumeration is not feasible. Fortunately, the solutions of a Boolean equation are related in such a way that condensed representations for S are readily constructed. One such representation is an interval (or intervals) defined by lower and upper bounds. Another representation is by means of a formula (or formulas) involving arbitrary parameters. The following theorem specifies such representations for the solutions of a single-variable Boolean equation.

Theorem 6.2.1 *Let $f: B \longrightarrow B$ be a Boolean function for which the equation $f(x) = 0$ is consistent, and let $S = \{x \mid f(x) = 0\}$ be the set of its solutions. Define subsets of B as follows:*

$$\begin{aligned} \text{(a)} \quad I &= \{x \mid f(0) \leq x \leq f'(1)\} \\ \text{(b)} \quad P &= \{f(0) + pf'(1) \mid p \in B\} \ . \end{aligned}$$

Then $I = P = S$.

Proof.   For notational simplicity we write f_0 and f_1, respectively, in place of $f(0)$ and $f(1)$. The equivalence of $f_0 \leq x \leq f_1'$ and $f(x) = 0$ (Proposition 4.1.1) implies that $I = \{x \mid f(x) = 0\} = S$. To prove that $P = S$, we first show that $P \subseteq S$:

$$f(f_0 + pf_1') = (f_0 + pf_1')'f_0 + (f_0 + pf_1')f_1 = f_0f_1 = ECON(f, \{x\}) = 0 \ .$$

Thus $P \subseteq S$. To verify that $S \subseteq P$, we show that for all a in B, the implication $[a \in S \implies a \in P]$ holds, i.e., that for any element $a \in B$ there is an element $p \in B$ such that

$$f(a) = 0 \quad \implies \quad f_0 + pf_1' = a \ .$$

Noting that $f(a)$ has the expanded form $a'f_0 + af_1$, applying Theorem 4.4.1 (the Extended Verification Theorem), and doing some computing, we reduce the foregoing implication to an equivalent equation, viz.,

$$f_0'f_1'[p \oplus a] = 0 \ ,$$

which is satisfied for $p = a$. Hence $S \subseteq P$. $\square$

Theorem 6.2.1 shows that the set of solutions of $f(x) = 0$, if not empty, may be expressed either as an *interval*,

$$f(0) \le x \le f'(1) , \tag{6.6}$$

or as a *parametric formula*,

$$x = f(0) + pf'(1) , \tag{6.7}$$

where the symbol p represents an *arbitrary parameter*, i.e., a freely-chosen member of B. Thus each of the two forms (6.6) and (6.7) is a general solution of $f(x) = 0$. These forms are condensed, relatively easy to produce, and provide a basis for enumerating, if necessary, the entire set of solutions. If the consistency-condition (6.4) is not satisfied identically, it should be stated as part of the solution.

Example 6.2.1 Let us find a general solution, having the interval-based form (6.6), of the equation

$$ax = b , \tag{6.8}$$

where a and b are fixed elements of a Boolean algebra B. Equation (6.8) reduces to the equivalent equation

$$ab'x + a'b + bx' = 0 . \tag{6.9}$$

The general solution (6.6) therefore takes the form

$$a'b + b \le x \le (ab' + a'b)' . \tag{6.10}$$

Interval (6.10) constitutes a general solution of (6.9) only if solutions exist, i.e., only if (6.9) is consistent. To determine the consistency-condition, given by equation (6.4), we calculate the conjunctive eliminant, ECON(f,{x}):

$$
\begin{aligned}
ECON(f, \{x\}) &= f(0)f(1) \\
&= (a'b + b)(ab' + a'b) .
\end{aligned}
$$

Thus equation (6.8) is consistent if and only if the condition

$$a'b = 0$$

is satisfied; this condition should accompany (6.10) as a complete statement of the general solution.

It often happens that the the consistency-condition can be used to simplify the form of a general solution. For example, the implication

$$a'b = 0 \implies a'b' + ab = a' + b$$

enables the general solution (6.10) to be written in the simpler form

$$a'b \ = \ 0 \tag{6.11}$$

$$b \ \leq \ x \ \leq \ a' + b \,. \tag{6.12}$$

The system (6.11, 6.12) is equivalent to the original equation, (6.8); however (6.12) alone is not equivalent to (6.8). Thus the consistency-condition is an essential part of the general solution (6.11, 6.12). The solution (6.10), however, is equivalent to (6.8) and therefore implies its own consistency-condition, i.e., $(6.8) \implies b \leq (a'b' + ab) \implies a'b = 0$. It is good practice in any case to state the consistency-condition as an explicit part of a general solution.

Example 6.2.2 Let us now construct a parametric general solution of equation (6.8), making use of formula (6.7). From the equivalent equation (6.9) we derive the discriminants $f(0) = b$ and $f(1) = ab' + a'b$; thus a general solution of (6.8) is

$$x = b + p(a'b' + ab) \,,$$

i.e.,

$$x = b + pa' \,,$$

with consistency-condition $a'b = 0$.

6.3   Subsumptive General Solutions

In this section we extend the interval-based general solution (6.6) to apply to Boolean equations having more than one unknown. We omit proofs, which are given in Brown & Rudeanu [28]. Let us consider an n-variable Boolean system of the form

$$
\begin{aligned}
s_0 \ &\leq \ 0 \\
s_1 \ &\leq \ x_1 \ \leq \ t_1 \\
s_2(x_1) \ &\leq \ x_2 \ \leq \ t_2(x_1) \\
s_3(x_1, x_2) \ &\leq \ x_3 \ \leq \ t_3(x_1, x_2) \\
&\vdots \\
s_n(x_1, \ldots, x_{n-1}) \ &\leq \ x_n \ \leq \ t_n(x_1, \ldots, x_{n-1})
\end{aligned}
\tag{6.13}
$$

where s_0, s_1, and t_1 are constants (elements of B) and the remaining s_i and t_i are Boolean functions having the indicated number of arguments. We say that the system (6.13) is a *subsumptive general solution* of the n-variable Boolean equation

$$f(x_1, x_2, \ldots, x_n) = 0 \qquad (6.14)$$

if $s_0 \leq 0$ (i.e., $s_0 = 0$) is the consistency-condition of (6.14) and if, provided (6.14) is consistent, every particular solution $(a_1, \ldots, a_n)$ of (6.14), and nothing else, is generated by the following procedure:

1)   Select a_1 in the range $s_1 \leq x \leq t_1$
2)   Select a_2 in the range $s_2(a_1) \leq x \leq t_2(a_1)$

$\vdots$

n)   Select a_n in the range $s_n(a_1, \ldots, a_{n-1}) \leq x \leq t_n(a_1, \ldots, a_{n-1})$.

This procedure enables all of the solutions of (6.14) to be enumerated as a tree. The form of the tree (but not the set of particular solutions it represents) depends on the sequence in which argument-values are generated. The argument-sequence $x_1, x_2, \ldots, x_n$, which is explicit in (6.13), will henceforth be assumed.

Example 6.3.1 A subsumptive general solution of the Boolean equation

$$yz + a'x' + a'z' + xy = 0$$

is

$$
\begin{array}{ccccc}
0 & = & 0 & & \\
a' & \leq & x & \leq & 1 \\
0 & \leq & y & \leq & x' \\
a' & \leq & z & \leq & y' .
\end{array}
$$

This general solution yields five particular solutions, which are listed in Table 6.1.

6.3.1   Successive Elimination

The classical method for producing a subsumptive general solution is by *successive elimination of variables*. This technique is part of the folklore of the subject; the first formal proof of its adequacy was apparently given by Rudeanu [171].

x	y	z
a'	0	a'
a'	0	1
a'	a	a'
1	0	a'
1	0	1

Table 6.1: Particular solutions of $yz + a'x' + a'z' + xy = 0$.

The idea behind successive elimination is to transform the problem of solving a single n-variable equation into that of solving n single-variable equations. The process begins with the elimination of x_n from (6.14). We call the conjunctive eliminant f_{n-1}; hence, the resultant is $f_{n-1}(x_1, \ldots, x_{n-1}) = 0$. The variable x_{n-1} is then eliminated from the latter equation, yielding the equation $f_{n-2}(x_1, \ldots, x_{n-2}) = 0$, the resultant of elimination of x_n and x_{n-1} from (6.14). This process is continued until the resultants $f_1(x_1)$ and, finally, f_0 are produced. The latter equation, the resultant of elimination of all variables from (6.14), is the necessary and sufficient condition for (6.14) to be consistent, i.e., solvable. If that condition is satisfied, then the single-variable equation $f_1(x_1) = 0$ is solved for x_1. For any solution $x_1 = a_1$, the single-variable equation $f_2(a_1, x_2) = 0$ is solved for x_2; for any solution $x_2 = a_2$ of the latter equation, the single-variable equation $f_3(a_1, a_2, x_3) = 0$ is solved for x_3; this process is continued, working back up the sequence of resultants until a solution $(a_1, a_2, \ldots, a_n)$ of (6.14) is achieved. We formalize this procedure in the following theorem.

Theorem 6.3.1 *Given the Boolean function $f \colon B^n \longrightarrow B$ define $n+1$ eliminants $f_0, f_1(x_1), f_2(x_1, x_2), \ldots, f_n(x_1, \ldots, x_n)$ of f by means of the recursion*

$$\text{(i)} \quad f_n = f \tag{6.15}$$
$$\text{(ii)} \quad f_{i-1} = ECON(f_i, \{x_i\}) \quad (i = n, n-1, \ldots, 1). \tag{6.16}$$

Then (6.13) is a subsumptive general solution of (6.14) provided the Boolean functions $s_0, s_1, \ldots, s_n, t_1, \ldots, t_n$ are defined by

$$s_0 = f_0 \tag{6.17}$$
$$s_i(x_1, \ldots, x_{i-1}) = f_i(x_1, \ldots, x_{i-1}, 0) \quad (i = 1, \ldots, n) \tag{6.18}$$
$$t_i(x_1, \ldots, x_{i-1}) = f_i'(x_1, \ldots, x_{i-1}, 1) \quad (i = 1, \ldots, n). \tag{6.19}$$

Example 6.3.2 Let us apply the method of successive elimination to find a general solution of the equation $f(x_1, x_2, x_3) = 0$, where f is represented by the formula

$$f = bx_1 + bx_2'x_3 + b'x_1'x_2 + a'x_2'x_3' + a'x_1x_2' + a'x_1'x_2 + ab'x_2x_3 . \quad (6.20)$$

The eliminants of f defined by the recursion (6.15, 6.16) are represented as follows:

$$
\begin{aligned}
f_3 &= \text{foregoing formula for } f \\
f_2 &= bx_1 + b'x_1'x_2 + a'x_1x_2' + a'x_1'x_2 + a'bx_2' \\
f_1 &= bx_1 + a'bx_1' \\
f_0 &= a'b .
\end{aligned}
$$

Thus the system

$$
\begin{aligned}
a'b &= 0 \\
a'b \le x_1 &\le b' \\
a'b + a'x_1 + bx_1 \le x_2 &\le b'x_1 + abx_1' \\
a'x_1' + bx_1 + b'x_1'x_2 + a'x_2' \le x_3 &\le b'x_1'x_2' + abx_1'x_2 + ab'x_2' + a'b'x_1x_2
\end{aligned}
\quad (6.21)
$$

is a subsumptive general solution of $f(x_1, x_2, x_3) = 0$. The SOP formulas in (6.21) may be simplified, as shown below, by introducing the condition $a'b = 0$ explicitly:

$$
\begin{aligned}
a'b &= 0 \\
0 \le x_1 &\le b' \\
a'x_1 + bx_1 \le x_2 &\le b'x_1 + bx_1' \\
a'x_1' + bx_1 + b'x_1'x_2 + a'x_2' \le x_3 &\le b'x_1'x_2' + bx_1'x_2 + ab'x_2' + a'x_1x_2 .
\end{aligned}
\quad (6.22)
$$

6.3.2 Deriving Eliminants from Maps

The eliminants $f_0, f_1, \ldots, f_n$ are readily derived if $f(x_1, \ldots, x_n)$ is represented by a 2^n-row map, each row of which represents $f(a_1, \ldots, a_n)$, where $(a_1, \ldots, a_n)$ is one of the elements of $\{0, 1\}^n$. The recursion (6.15, 6.16) implies that each row of the 2^{i-1}-row map representing f_{i-1} is the result of intersecting a pair of rows of the 2^i-row map representing f_i. The result is particularly convenient if f is represented by a Karnaugh map [99], in which the rows and columns are arranged according to a reflected Gray code, or by a Marquand diagram [131], in which the rows and columns are arranged in natural binary order. For either of these representations, the map representing f_{i-1} is constructed from the map representing f_i by intersecting successive pairs of rows, beginning with the top pair. Karnaugh maps are easier to construct and read than are Marquand diagrams; however, the rules for using Marquand diagrams to solve Boolean equations are easier to state than are those for using Karnaugh maps. We therefore specialize to Marquand diagrams, inasmuch as the Marquand-rules are readily converted in practice to Karnaugh-rules.

Example 6.3.3 The function given in Example 6.3.2 is represented in Figure 6.1 as Marquand diagram f_3. Each row of diagram f_2 is formed by intersecting (element-by-element ANDing) a pair of rows of f_3; thus row 00 of diagram f_2 is the intersection of rows 000 and 001 of diagram f_3. Diagram f_1 is constructed in a similar way from diagram f_2, and so on.

6.3.3   Recurrent Covers and Subsumptive Solutions

The general solution produced by successive elimination is only one among many subsumptive general solutions typically possessed by a Boolean equation. The eliminants $f_0, f_1, \ldots, f_n$ defined by (6.15, 6.16) may be used, however, to generate the full class of subsumptive general solutions; knowing that class, we may select a general solution best suited to our purposes.

We show in Theorem 6.3.4 that each subsumptive general solution of the Boolean equation (6.14) is associated with a sequence $(g_0, g_1, \ldots, g_n)$ of Boolean functions. We say that such a sequence is *recurrent* in case $g_0 \in B$ and also that $g_i \colon B^i \longrightarrow B$ is a Boolean function of the variables $x_1, x_2, \ldots, x_i$ satisfying the condition

$$ECON(g_i, \{x_i\}) \leq \sum_{j=0}^{i-1} g_j(x_1, \ldots, x_j)$$

for $i = 1, 2, \ldots, n$. The sequence $(g_0, g_1, \ldots, g_n)$ will be called a *recurrent cover* of an n-variable Boolean function f in case the sequence is recurrent and also

$$f = \sum_{j=0}^{n} g_j .$$

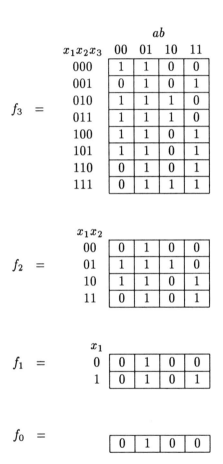

$f_3 =$

	ab			
$x_1x_2x_3$	00	01	10	11
000	1	1	0	0
001	0	1	0	1
010	1	1	1	0
011	1	1	1	0
100	1	1	0	1
101	1	1	0	1
110	0	1	0	1
111	0	1	1	1

$f_2 =$

x_1x_2				
00	0	1	0	0
01	1	1	1	0
10	1	1	0	1
11	0	1	0	1

$f_1 =$

x_1				
0	0	1	0	0
1	0	1	0	1

$f_0 =$

0	1	0	0

Figure 6.1: Marquand diagrams for the eliminants of (6.20).

Theorem 6.3.2 *Let $f_0, f_1, \ldots, f_n$ be the eliminants, defined by (6.15, 6.16), of an n-variable function f. Then $(g_0, g_1, \ldots, g_n)$ is a recurrent cover of f if and only if the conditions*

$$
\begin{aligned}
f_0 &= g_0 \\
f_0' f_1 &\leq g_1 \leq f_1 \\
f_1' f_2 &\leq g_2 \leq f_2 \\
&\vdots \\
f_{n-1}' f_n &\leq g_n \leq f_n
\end{aligned}
$$

are satisfied.

Corollary 6.3.1 *The sequence $(f_0, f_1, \ldots, f_n)$ of eliminants defined by the system (6.15, 6.16) is a recurrent cover of f.*

   Given the eliminants $f_0, f_1, \ldots, f_n$ of an n-variable Boolean function f, Theorem 6.3.2 expresses the set of recurrent covers of f by a system of intervals, i.e., of "incompletely specified" functions. Thus well-known methods of minimization may be used to find a recurrent cover expressed by the simplest possible SOP formulas. The next theorem shows, however, that a recurrent cover of f may be constructed directly from the prime implicants of f.

Theorem 6.3.3 *Given an n-variable Boolean function f, a recurrent cover $(g_0, g_1, \ldots, g_n)$ of f is given by the prescriptions*

$$
\begin{aligned}
g_0 &= \sum (\text{prime implicants of } f \text{ not involving} \\
&\qquad \text{any of the arguments } x_1, \ldots, x_n) \qquad\qquad (6.23) \\
g_i &= \sum (\text{prime implicants of } f \text{ involving } x_i \\
&\qquad \text{but not involving any of } x_{i+1}, \ldots, x_n) \\
&\qquad\qquad\qquad\qquad\qquad\qquad (i = 1, 2, \ldots, n) . \qquad (6.24)
\end{aligned}
$$

   It is convenient in practice to produce the g-sequence specified by (6.23, 6.23) in reverse order, i.e., beginning with g_n rather than with g_0. Let F be the set of prime implicants of f, and denote by G_i the set of terms comprised by $g_i(i = 0, \ldots, n)$. Then the sets $G_n, G_{n-1}, \ldots, G_0$ (which constitute a partition of F) are generated by procedure shown in Figure 6.2, in which T is a subset of F.

```
1   begin
2       T := F
3       for i = n downto 1 do
4           begin
5               G_i := {terms in T involving x_i}
6               T := T - G_i
7           end
8       G_0 := T
9   end
```

Figure 6.2: Procedure to generate a recurrent cover of f from the prime implicants of f.

Example 6.3.4 Let us apply the procedure of Figure 6.2 to produce a recurrent cover of the Boolean function f discussed in Example 6.3.2. The Blake canonical form of f, i.e., the disjunction of its prime implicants, is

$$BCF(f) = bx_1 + bx_2'x_3 + b'x_1'x_2 + a'x_2'x_3' + a'x_1x_2' + a'x_1'x_2 + ab'x_2x_3 + a'b + ax_1x_2x_3 + a'x_1'x_3' .$$

The procedure of Figure 6.2 enables us to read off the components of a recurrent cover by inspection of BCF(f):

$$
\begin{aligned}
g_3 &= bx_2'x_3 + a'x_2'x_3' + ab'x_2x_3 + ax_1x_2x_3 + a'x_1'x_3' \\
g_2 &= b'x_1'x_2 + a'x_1x_2' + a'x_1'x_2 \\
g_1 &= bx_1 \\
g_0 &= a'b .
\end{aligned}
\tag{6.25}
$$

The following theorem shows the intimate relation between recurrent covers and subsumptive general solutions.

Theorem 6.3.4 *The system (6.13) is a subsumptive general solution of equation* $f(x_1, x_2, \ldots, x_n) = 0$ *if and only if* $s_0, s_1, \ldots, s_n, t_1, \ldots, t_n$ *are Boolean functions given by*

$$
\begin{aligned}
s_0 &= g_0 \\
s_i(x_1, \ldots, x_{i-1}) &= g_i(x_1, \ldots, x_{i-1}, 0) \quad (i = 1, \ldots, n) \\
t_i(x_1, \ldots, x_{i-1}) &= g_i'(x_1, \ldots, x_{i-1}, 1) \quad (i = 1, \ldots, n)
\end{aligned}
$$

where $(g_0, g_1, \ldots, g_n)$ *is a recurrent cover of* f.

Example 6.3.5 The method of successive elimination was applied in Example 6.3.2 to construct a general solution of the equation $f(x_1, x_2, x_3) = 0$, the function f being specified by (6.20). Let us now apply Theorem 6.3.4 to construct another general solution of the same equation, based on the recurrent cover (6.25).

$$
\begin{aligned}
a'b &= 0 \\
0 \leq x_1 &\leq b' \\
a'x_1 \leq x_2 &\leq ab + x_1 \\
a'x_1' + a'x_2' \leq x_3 &\leq b'x_2' + a'x_2 + bx_1'x_2
\end{aligned}
\tag{6.26}
$$

The formulas in the general solution (6.26) are clearly simpler than those in either (6.21) or (6.22):

6.3.4   Simplified Subsumptive Solutions

The method of successive eliminations tends to produce unnecessarily complex formulas for the s_i and t_i in a subsumptive general solution, even if the consistency-condition $f_0 = 0$ is introduced, as in (6.22), for purposes of simplification. Such complex formulas mask the nature of the solutions and complicate the task of enumerating particular solutions.

We call subsumptive solution (6.13) *simplified* in case each of the functions $s_0, s_1, \ldots, s_n, t_1, \ldots, t_n$ is expressed as a simplified SOP formula. The following theorem establishes that each of these functions (like each of the g_i in a recurrent cover of f) is defined by an interval based on the eliminants $f_0, f_1, \ldots, f_n$. Thus we may apply standard procedures for minimizing the complexity of SOP formulas.

Theorem 6.3.5 *Let* $f_0, f_1, \ldots, f_n$ *be the eliminants, defined by (6.15, 6.16), of an n-variable Boolean function f. Then (6.13) is a subsumptive general solution of* $f(x_1, x_2, \ldots, x_n) = 0$ *if and only if*

$$
s_0 = f_0
\tag{6.27}
$$

and the conditions

$$
f_i(0)f_i'(1) \leq s_i \leq f_i(0)
\tag{6.28}
$$
$$
f_i'(1) \leq t_i \leq f_i(0) + f_i'(1)
\tag{6.29}
$$

are satisfied for $i = 1, 2, \ldots, n$, where $f_i(0)$ and $f_i(1)$ denote, respectively, $f_i(x_1, \ldots, x_{i-1}, 0)$ and $f_i(x_1, \ldots, x_{i-1}, 1)$.

6.3.5 Simplification via Marquand Diagrams

If $f_0, f_1, \ldots, f_n$ are represented by Marquand diagrams, as in Example 6.3.3, then diagrams representing all possible values of $s_0, s_1, \ldots, s_n, t_1, \ldots, t_n$ are readily derived by use of relations (6.27, 6.28, 6.29). Each column of the diagrams for $s_i(x_1, \ldots, x_{i-1})$ and $t_i(x_1, \ldots, x_{i-1})$ is derived from the corresponding column of $f_i(x_1, \ldots, x_i)$. The element in row k ($k = 0, 1, \ldots, 2^{i-1} - 1$) for any column of the s_i and t_i diagrams is related as shown in Table 6.2 to the elements in rows $2k$ and $2k + 1$ of the diagram for f_i.

row $2k$ of f_i	0	0	1	1
row $2k + 1$ of f_i	0	1	0	1
row k of s_i	0	0	1	X
row k of t_i	1	0	1	X

Table 6.2: Diagram-entries for s_i and t_i in terms of diagram-entries for f_i.

Example 6.3.6 The eliminants of function (6.20) are represented by Marquand diagrams in Example 6.3.3. Those diagrams, together with the corresponding diagrams for s_i and t_i, are exhibited in Figure 6.3. The Marquand diagrams for s_i and t_i specify the set of all possible subsumptive general solutions for the Boolean equation of Example 6.3.2. A simplified member of that set is

$$a'b = 0$$
$$0 \leq x_1 \leq b'$$
$$a'x_1 \leq x_2 \leq b + x_1$$
$$a'x_1' \leq x_3 \leq a' + b'x_2' + bx_2$$

The form of this general solution should be compared with that of the general solutions produced in Examples 6.3.2 and 6.3.5.

6.4 Parametric General Solutions

Formula (6.7) expresses a parametric general solution of the single-variable Boolean equation $f(x) = 0$. We now consider parametric general solutions of n-variable Boolean equations.

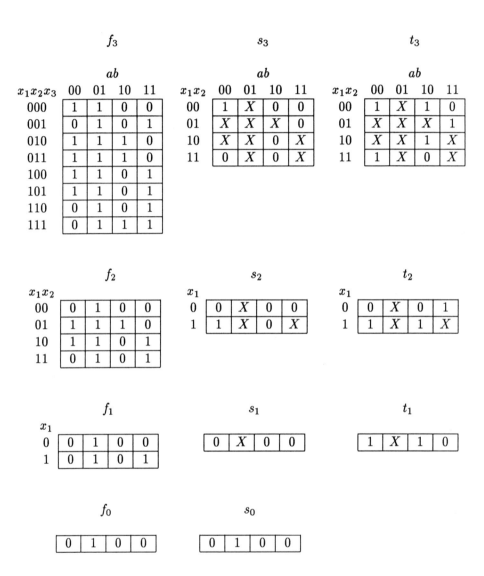

Figure 6.3: Marquand diagrams associated with (6.20).

Let X, G, and P denote, respectively, the vectors $(x_1, \ldots, x_n)$, $(g_1, \ldots, g_n)$, and $(p_1, \ldots, p_k)$, where $g_1, \ldots, g_n$ are k-variable Boolean functions and where the symbols $p_1, \ldots, p_k$ designate *arbitrary parameters*, i.e., freely-chosen elements of B. Then a *parametric general solution* of the Boolean equation

$$f(X) = 0 \qquad (6.30)$$

is a system

$$0 = g_0 \qquad (6.31)$$
$$X = G(P) \qquad (6.32)$$

such that the conditions

$$g_0 = ECON(f, X) \qquad (6.33)$$
$$f(G(P)) = ECON(f, X) \quad \forall P \in B^k \qquad (6.34)$$
$$f(A) = ECON(f, X) \implies \exists P \in B^k \text{ such that } G(P) = A \qquad (6.35)$$

are satisfied.

Condition (6.33) specifies that equation (6.31) is to be the consistency-condition for (6.30). Suppose condition (6.33) to be satisfied. Then condition (6.34) demands that $G(P)$ generate nothing but solutions of (6.30) for all values of the parameter-vector P; condition (6.35), on the other hand, demands that $G(P)$ generate all solutions of (6.30). If (6.30) is consistent, therefore, conditions (6.34) and (6.35) taken together demand that the set of all of its solutions—and nothing else—be generated by $G(P)$ as P is assigned values on B^k.

6.4.1 Successive Elimination

The method of successive elimination of variables, which we applied earlier to find subsumptive general solutions, can also be used to find parametric general solutions. To acquaint ourselves with the main features of the procedure, let us solve the 3-variable equation

$$f(x_1, x_2, x_3) = 0 . \qquad (6.36)$$

We begin, as before, by calculating the eliminants f_0, f_1, f_2, f_3:

$$
\begin{aligned}
f_3(x_1, x_2, x_3) &= ECON(f, \emptyset) &= f \\
f_2(x_1, x_2) &= ECON(f_3, \{x_3\}) &= ECON(f, \{x_3\}) \\
f_1(x_1) &= ECON(f_2, \{x_2\}) &= ECON(f, \{x_2, x_3\}) \\
f_0 &= ECON(f_1, \{x_1\}) &= ECON(f, \{x_1, x_2, x_3\})
\end{aligned}
$$

The consistency-condition for equation (6.36) is $0 = g_0$, where $g_0 = f_0$. Equation $f_1(x_1) = 0$ is the resultant of elimination of x_2 and x_3 from (6.36). This is a single-variable equation stating all that is known about x_1 in the absence of knowledge concerning x_2 and x_3; hence, we may employ the parametric formula (6.7) to express the set of allowable values of x_1, i.e.,

$$x_1 = g_1(p_1) = f_1(0) + p_1 f_1'(1) . \qquad (6.37)$$

Equation $f_2(x_1, x_2) = 0$ is the resultant of elimination of x_3 from (6.36); it therefore expresses all that is known about x_1 and x_2 in the absence of knowledge concerning x_3. We substitute $g_1(p_1)$ for x_1, re-expressing this resultant by the single-variable equation $f_2(g_1(p_1), x_2) = 0$; a general solution of the latter equation is

$$\begin{aligned} x_2 &= g_2(p_1, p_2) \\ &= f_2(g_1(p_1), 0) + p_2 f_2'(g_1(p_1), 1) . \end{aligned} \qquad (6.38)$$

The final step in this process is to substitute $g_1(p_1)$ and $g_2(p_1, p_2)$ for x_1 and x_2, respectively, in (6.36). The result, $f_3(g_1(p_1), g_2(p_1, p_2), x_3) = 0$, is a single-variable equation in x_3 whose general solution is expressed by the parametric formula

$$\begin{aligned} x_3 &= g_3(p_1, p_2, p_3) \\ &= f_3(g_1(p_1), g_2(p_1, p_2), 0) + p_3 f_3'(g_1(p_1), g_2(p_1, p_2), 1) . \end{aligned} \qquad (6.39)$$

The system

$$\begin{aligned} 0 &= g_0 \\ x_1 &= g_1(p_1) \\ x_2 &= g_2(p_1, p_2) \\ x_3 &= g_3(p_1, p_2, p_3) \end{aligned}$$

therefore constitutes a parametric general solution of (6.36), the functions g_0, g_1, g_2, and g_3 being defined by (6.37, 6.38, 6.39).

Example 6.4.1 The *RST flip-flop* is defined by the equations

$$\begin{aligned} Y &= S + yR'T' + y'T \qquad &(6.40) \\ 0 &= RS + RT + ST , \qquad &(6.41) \end{aligned}$$

where y is the present state of the flip-flop, Y is its next state, and R, S, and T are the "reset", "set", and "toggle" inputs to the flip-flop. Equation (6.40) is the *characteristic equation* [155] of the flip-flop and (6.41) is an input-constraint specifying that no more than one input may be at logical 1 at any time. To design input-logic for this flip-flop, it is necessary to solve the system (6.40, 6.41) for R, S, and T (which we regard as variables) in terms of y and Y (which we regard as elements of B). We begin by reducing the system (6.40, 6.41) to the single equivalent equation

$$f(R, S, T) = 0 ,$$

where f is defined by

$$f = Y'S + y'Y'T + yY'R'T' + y'YS'T' + yYRS' + yYS'T \\ + RS + RT + ST . \tag{6.42}$$

Let us now employ successive elimination to form a parametric general solution. To make direct use of the results given above concerning the formation of such a solution, we re-name variables as follows: $x_1 = R$, $x_2 = S$, and $x_3 = T$. The RST flip-flop is therefore defined by $f(x_1, x_2, x_3) = 0$, where f is given by

$$f = Y'x_2 + y'Y'x_3 + yY'x_1'x_3' + y'Yx_2'x_3' + yYx_1x_2' + yYx_2'x_3 \\ + x_1x_2 + x_1x_3 + x_2x_3 . \tag{6.43}$$

Accordingly, the eliminants are

$$\begin{aligned} f_3(x_1, x_2, x_3) &= f &= \text{expression (6.43)} \\ f_2(x_1, x_2) &= ECON(f_3, \{x_3\}) &= Yx_1 + Y'x_2 \\ f_1(x_1) &= ECON(f_2, \{x_2\}) &= Yx_1 \\ f_0 &= ECON(f_1, \{x_1\}) &= 0 . \end{aligned}$$

Substituting the foregoing results in the parametric expressions (6.37, 6.38, 6.39) we arrive at the general solution

$$\begin{aligned} 0 &= 0 \\ x_1 &= 0 + p_1Y' \\ x_2 &= 0 + p_2Y \\ x_3 &= [p_1'yY' + p_2'y'Y] + p_3[p_1Y' + p_2Y + yY + y'Y']' . \end{aligned}$$

After replacement of x_1, x_2, and x_3, by R, S, and T, respectively, and simplification of the result, the foregoing general solution takes the form

$$
\begin{aligned}
0 &= 0 \\
R &= p_1 Y' \\
S &= p_2 Y \\
T &= p_1' y Y' + p_2' y' Y \ .
\end{aligned}
\tag{6.44}
$$

The first equation in the system (6.44), $0 = 0$, signifies that the RST flip-flop equation is unconditionally consistent; for any combination of present-state and next-state, that is, the equation has a solution for R, S, and T. Particular solutions are obtained from (6.44) by assignment of particular values in the Boolean algebra B to the arbitrary parameters. The Boolean algebra B is not specified in the foregoing example, but includes as a subalgebra the free Boolean algebra $FB(y, Y)$. Suppose we make the assignment $p_1 = 1$, $p_2 = y$. The corresponding particular solution is $R = Y'$, $S = yY$, $T = y'Y$.

The method of successive elimination demonstrates that a parametric general solution of an n-variable Boolean equation need involve no more than n arbitrary parameters. In some cases, as shown by the foregoing example, fewer than n parameters suffice.

6.4.2   Parametric Solutions based on Recurrent Covers

As we found in discussing subsumptive general solutions, the sequence $f_0, f_1, \ldots, f_n$ of eliminants of f is a special form of recurrent cover of f. We also found that the eliminants of f constitute one of the more complex recurrent covers of f, and that simpler recurrent covers may be constructed in a systematic way; the procedure of Theorem 6.3.3, for example, develops a relatively simple recurrent cover of f directly from its prime implicants.

The method of successive elimination may be generalized directly to employ any recurrent cover of f as the basis for constructing a parametric general solution of $f(X) = 0$. We now present several results, without proof, concerning the use of recurrent covers for this purpose.

Our first observation is that if $(g_0, g_1, \ldots, g_n)$ is a recurrent cover of $f \colon B^n \longrightarrow B$, then the "triangular" system

$$0 = \hat{g}_0$$
$$x_1 = \hat{g}_1(p_1)$$
$$x_2 = \hat{g}_2(p_1, p_2)$$
$$\vdots$$
$$x_n = \hat{g}_n(p_1, p_2, \ldots, p_n)$$

is a parametric general solution of the Boolean equation $f(x_1, \ldots, x_n) = 0$ if the functions $\hat{g}_0, \hat{g}_1, \ldots, \hat{g}_n$ are given by the recursion

$$\hat{g}_0 = g_0 \tag{6.45}$$
$$\hat{g}_1(p_1) = g_1(0) + p_1 g_1'(1) \tag{6.46}$$
$$\hat{g}_i(p_1, \ldots, p_i) = g_i(g_1(p_1), \ldots, g_{i-1}(p_1, \ldots, p_{i-1}), 0) +$$
$$+ p_i g_i'(g_1(p_1), \ldots, g_{i-1}(p_1, \ldots, p_{i-1}), 1) \tag{6.47}$$
$$(i = 2, 3, \ldots, n) .$$

We note further that the functions $\hat{g}_0, \hat{g}_1(p_1), \hat{g}_2(p_1, p_2), \ldots, \hat{g}_n(p_1, \ldots, p_n)$ are independent of the recurrent cover $(g_0, g_1, \ldots, g_n)$ of f on which they are based, provided $f(X) = 0$ is consistent [28]. Thus all recurrent covers lead via the system (6.45, 6.46, 6.47) to the same general solution of a consistent Boolean equation.

Example 6.4.2 Let us form a parametric general solution of the Boolean equation

$$ax_1 + bx_2 = c , \tag{6.48}$$

which is equivalent to the equation $f(x_1, x_2) = 0$, where f is given by

$$f = ac'x_1 + bc'x_2 + a'b'c + a'cx_2' + b'cx_1' + cx_1'x_2' .$$

The foregoing expression is in Blake canonical form; hence, we may form a recurrent cover by inspection of its terms, using the procedure of Theorem 6.3.3:

$$g_0 = a'b'c$$
$$g_1(x_1) = ac'x_1 + b'cx_1'$$
$$g_2(x_1, x_2) = bc'x_2 + a'cx_2' + cx_1'x_2' .$$

Thus, applying the recursion (6.45, 6.46, 6.47),

$$
\begin{aligned}
\hat{g}_0 &= g_0 \\
&= a'b'c \\
\hat{g}_1(p_1) &= b'c + p_1[ac']' \\
&= b'c + p_1[a' + c] \\
\hat{g}_2(p_1, p_2) &= a'c + cg_1'(p_1) + p_2[bc']' \\
&= a'c + p_1'bc + p_2[b' + c] \,,
\end{aligned} \tag{6.49}
$$

from which we form the parametric general solution

$$
\begin{aligned}
0 &= a'b'c \\
x_1 &= b'c + p_1[a' + c] \\
x_2 &= a'c + p_1'bc + p_2[b' + c] \,.
\end{aligned}
$$

Example 6.4.3 The recurrent cover

$$
\begin{aligned}
g_3 &= bx_2'x_3 + a'x_2'x_3' + ab'x_2x_3 + ax_1x_2x_3 + a'x_1'x_3' \\
g_2 &= b'x_1'x_2 + a'x_1x_2' + a'x_1'x_2 \\
g_1 &= bx_1 \\
g_0 &= a'b
\end{aligned}
$$

was developed in Example 6.3.4 for the function f given in Example 6.3.2. Thus a parametric general solution of $f(x_1, x_2, x_3) = 0$ is

$$
\begin{aligned}
0 &= a'b \\
x_1 &= p_1 b' \\
x_2 &= a'b'p_1 + p_2[b'p_1 + ab] \\
x_3 &= p_1'a' + a'b + p_3[p_1'b' + p_2b + a' + p_2'b'] \,.
\end{aligned}
$$

We observe finally that if $f(X) = 0$ is consistent, the general solution defined by the recursion (6.45, 6.46, 6.47) is *reproductive*; that is, the equivalence

$$
f(X) = 0 \quad \Longleftrightarrow \quad X = \hat{G}(X) \tag{6.50}
$$

holds for all X in B^n. (See Rudeanu [172] for a full discussion of reproductive general solutions.)

Example 6.4.4 To illustrate the equivalence (6.50), let us assume that (6.48) is consistent and take at random one of its particular solutions:

$$x_1 = a'b' + b'c$$
$$x_2 = bc .$$

That this is a solution is verified by substituting in (6.48); the result,

$$a(a'b' + b'c) + b(bc) = c ,$$

becomes an identity provided that the consistency-condition, $0 = a'b'c$, is an identity. Thus the left side of (6.50) is satisfied. To verify the right side, we make the substitutions $p_1 = a'b' + b'c$ and $p_2 = bc$ in (6.49):

$$
\begin{aligned}
\hat{g}_1(a'b' + b'c) &= b'c + (a'b' + b'c)(a' + c) \\
&= a'b' + b'c \\
\hat{g}_2(a'b' + b'c, bc) &= a'c + (a'b' + b'c)'bc + bc(b' + c) \\
&= a'c + bc \\
&= bc \quad \text{(assuming consistency)}.
\end{aligned}
$$

6.4.3 Löwenheim's Formula

In cases where it is inconvenient to construct a general solution of a Boolean equation, it may be relatively simple to find a particular solution. The following theorem enables us to form a parametric general solution in a mechanical way from any particular solution.

Theorem 6.4.1 (Löwenheim [124]) *Let $U = (u_1, \ldots, u_n)$ be a particular solution of a consistent Boolean equation $f(x) = 0$, and let $P = (p_1, \ldots, p_n)$ be an n-tuple of arbitrary parameters. Then the system (6.31, 6.32) is a parametric general solution provided the vector $\hat{G}(P) = (\hat{g}_1(P), \ldots, \hat{g}_n(P))$ is defined by*

$$\hat{G}(P) = U f(P) + P f'(P) . \tag{6.51}$$

Proof. Condition (6.33) of the definition of a parametric general solution is verified because $f(X) = 0$ is assumed to be consistent, i.e., $ECON(f, X) = 0$ identically. Condition (6.34) is verified, for $k = n$, by Lemma 5.4.1. To verify (6.35), let us suppose that $A \in B^n$ is any particular solution of $f(X) = 0$. Choosing $P = A$,

$$\hat{G}(P) = U f(A) + A f'(A) = U \cdot 0 + A \cdot 1 = A .$$

□

Example 6.4.5 The RST flip-flop is characterized by the Boolean equation $f(R, S, T) = 0$, where f is given by (6.42). An easily-obtained particular solution of the RST equation is

$$
\begin{aligned}
R &= Y' \\
S &= Y \\
T &= 0 \, .
\end{aligned}
$$

The value of n for this example is 3; thus, the parameter-vector P has three components, which we call $p, q,$ and r for simplicity. Formula (6.51) therefore becomes

$$
\begin{bmatrix} \hat{g}_1(P) \\ \hat{g}_2(P) \\ \hat{g}_3(P) \end{bmatrix} = \begin{bmatrix} Y' \\ Y \\ 0 \end{bmatrix} f(p, q, r) + \begin{bmatrix} p \\ q \\ r \end{bmatrix} f'(p, q, r)
$$

The resulting parametric general solution is found after some calculation to be

$$
\begin{aligned}
0 &= 0 \\
R &= (p + q + r'y + ry')Y' \\
S &= (p + q + ry + r'y')Y \\
T &= p'q'r(y'Y + yY') \, .
\end{aligned}
$$

Löwenheim's formula generates an n-parameter solution of an n-variable Boolean equation. As Example 6.4.1 shows, however, fewer than n parameters will often suffice. As a further example, a two-parameter general solution of the RST flip-flop equation can be formed from the solution given in Example 6.4.5 by replacing the parameter-sum $p + q$ by the single parameter p, yielding

$$
\begin{aligned}
0 &= 0 \\
R &= (p + r'y + ry')Y' \\
S &= (p + ry + r'y')Y \\
T &= p'r(y'Y + yY') \, .
\end{aligned}
$$

It is possible, although the details will not be treated here, to generate least-parameter general solutions in a systematic way [23]; the Boolean equations characterizing all of the useful flip-flops (RS, JK, D, RST), for example, can be solved using no more than one parameter [24]. A one-parameter general solution of the RST flip-flop equation, for example, is

$$0 = 0$$
$$R = pY'$$
$$S = pY$$
$$T = p'(y'Y + yY').$$

Exercises

1. Prove that $f(x) = x$ is equivalent to $f(0) \leq x \leq f(1)$.

2. Give necessary and sufficient conditions on $f(0)$ and $f(1)$ in order for the equation $f(x) = 0$ to have a unique solution.

3. Let f be symmetrical with respect to x and x', i.e., let $f(x) = f(x')$. Discuss the solutions of $f(x) = 0$.

4. Construct simplified general solutions of the following systems of equations; state the consistency condition in each case.

   (a) $bx' + a = x$
         $a'x + b' = b'x$

   (b) $abx' + by + a'x = 1$

   (c) $x + y' = 1$
         $axy' + by = b'x'$
         $xy = a$

5. Consider the Boolean equation $f(x_1, x_2, x_3) = 0$, where

$$f = a_2x_1 + a_2x_2'x_3 + a_2'x_1'x_2 + a_1x_2'x_3'$$
$$+a_1x_1x_2' + a_1x_1'x_2 + a_1'a_2'x_2x_3 .$$

   (a) Find $BCF(f)$.
   (b) Construct a general solution using successive eliminations.
   (c) Construct a general solution using $BCF(f)$ directly.
   (d) Construct a general solution using maps.

6. How may the number of solutions of the Boolean equation $f(x) = 0$ be determined from a Karnaugh map representing f?

7. The RST flip-flop is defined by the equations

$$
\begin{aligned}
Y &= S + yR'T' + y'T \quad \text{(characteristic equation)} \\
0 &= RS + RT + ST \quad \text{(input constraint)}
\end{aligned}
$$

where y and Y are the present state and next state, respectively, of the flip-flop. The problem in the design of excitation-logic is to find economical solutions for R, S, and T in terms of Y (a given function of input and state variables) and y. Write a general solution for R, S, and T in the form

$$
\begin{aligned}
\alpha(y, Y) &= 0 \\
\alpha(y, Y) &\leq R \leq \beta(y, Y) \\
\alpha(y, Y, R) &\leq S \leq \beta(y, Y, R) \\
\alpha(y, Y, R, S) &\leq T \leq \beta(y, Y, R, S)
\end{aligned}
$$

Simplify the α's and β's.

8. The JK, SR, D and T flip-flops are defined by

JK: $Y = y'J + yK'$

SR: $Y = S + yR'$ and $SR = 0$

D: $Y = D$

T: $Y = y \oplus T$

Write general solutions and simplified particular solutions for

(a) S and R in terms of J, K, and y;

(b) D in terms of J, K, and y;

(c) J and K in terms of T and y;

(d) T in terms of S, R, and y.

9. Let the equation $a'x_1 + a'x_2 + ax_1' = 1$ be defined on the Boolean algebra $B = \{0, 1, a', a\}$.

(a) Display a general solution in the form

$$\alpha_1 \quad \leq \quad x_1 \quad \leq \quad \beta_1$$
$$\alpha_2(x_1) \quad \leq \quad x_2 \quad \leq \quad \beta_2(x_1)$$

Express the α and β functions by simplified formulas.

(b) Use your general solution in a systematic way to enumerate all particular solutions (x_1, x_2).

10. Given the Boolean algebra $\{0, 1, a, a'\}$, find the particular solutions of the Boolean equation $y'z' + a'y + ax' = 0$.

11. (a) We are given Boolean functions $f: B \longrightarrow B$ and $\phi: B \longrightarrow B$. Describe the steps needed to prove that $\phi(t)$ is a parametric general solution of the Boolean equation

$$f(x) = 0 .$$

(b) Suppose that $B = \{0, a', a, 1\}$. Show that the system

$$\phi_1(t) \quad = \quad at$$
$$\phi_2(t) \quad = \quad a't$$

is a parametric general solution of the Boolean equation

$$a'x_1 + ax_2 = 0 .$$

Your demonstration should not entail enumeration of all particular solutions.

12. (Boole [13], Chapter IX) Let us assume wealth to be defined as follows: "Wealth consists of things transferable, limited in supply, and either productive of pleasure or preventive of pain." Thus wealth is defined by the class-equation

$$w = st(p + r) ,$$

where w stands for wealth, s for things limited in supply, t for things transferable, p for things productive of pleasure, and r for things preventive of pain. State a general solution for the things transferable and productive of pleasure, i.e., the class pt, in terms of the classes r, s, and w. State a condition on the latter three classes that is necessary and sufficient for the existence of your solution.

Chapter 7

Functional Deduction

The central process of Boolean reasoning is the extraction of a derived system from a given Boolean system (*cf.* Section 4.2). The derived system may be categorized as

- functional (of the form $X = F(Y)$) or general (*i.e.*, not necessarily functional).
- antecedent or consequent.

The primary reasoning tactic employed by Boole and later nineteenth-century logicians was to solve a logical equation for certain of its arguments in terms of others. Boole's approach may thus be classified as functional and antecedent. We discuss such reasoning in Chapter 6.

Methods of general Boolean reasoning, both antecedent and consequent, were devised by Poretsky [159] at the end of the nineteenth century; these methods, however, entail computation of impractical complexity. A practical approach to general consequent reasoning in Boolean algebras was given by Blake [10] in 1937. Blake's "syllogistic" approach, which we describe in Chapter 5, is the Boolean precursor of Robinson's resolution principle [168] in the predicate calculus.

The object of the present chapter is to discuss *functional consequents* of a Boolean equation. Unlike functional antecedents (*i.e.*, solutions), which have been the object of much study since Boole's time, functional consequents seem to have received little attention.

We proceed throughout this chapter from a given consistent Boolean equation of the form

$$f(x_1, x_2, \ldots, x_n) = 1 \, . \tag{7.1}$$

(The 1-normal form (7.1) is chosen, in place of the 0-normal form used earlier in this book, for computational convenience.) Our objective is to find consequents of (7.1) having the form

$$x_1 = g(x_2, \ldots, x_n) \, .$$

Such consequents were called "consequent solutions" by Ledley [115, 118], who discussed their utility in circuit-design and proposed matrix methods for their construction.

We present criteria in Section 7.1 for determining the variables that are functionally deducible from equation (7.1). The associated functions are defined as intervals, $i.e.$, as incompletely-specified functions, enabling the methods of switching theory to be employed for minimization. In Section 7.2 we consider the following problem: given that a variable, u, is functionally deducible from (7.1), what are the minimal sets of variables from which the value of u may be computed?

7.1 Functionally Deducible Arguments

Assume as before that (7.1) is consistent, and consider a Boolean system having the form

$$
\begin{aligned}
u_1 &= g_1(V) \\
&\vdots \\
u_r &= g_r(V)
\end{aligned}
\tag{7.2}
$$

where $U = (u_1, \ldots, u_r)$ and $V = (v_1, \ldots, v_s)$ are disjoint subvectors of $X = (x_1, \ldots, x_n)$ and where $g_1, \ldots, g_r : \mathbf{B}^s \longrightarrow \mathbf{B}$ are Boolean functions. If (7.1) $\Longrightarrow$ (7.2), we say that (7.2) is a $functional\ consequent$ of (7.1), and that each of the arguments $u_1, \ldots, u_r$ is $functionally\ deducible$ from (7.1).

We confine our study of functionally deducible arguments to the case $r = 1$. Given a partition $\{\{u\}, V\}$ of $\{x_1, \ldots, x_n\}$, therefore, we define u to be functionally deducible from (7.1) in case there is a Boolean function $g : \mathbf{B}^{n-1} \longrightarrow \mathbf{B}$ such that the equation

$$u = g(V) \tag{7.3}$$

is a consequent of (7.1).

Theorem 7.1.1 *The following statements are equivalent:*

    (i) u is functionally deducible from (7.1).

    (ii) $EDIS(f', \{u\}) = 1$.

    (iii) $ECON(f, \{u\}) = 0$.

    (iv) u or u' appears in every term of $BCF(f)$.

Proof.

(i) $\Longleftrightarrow$ (ii) $\Longleftrightarrow$ (iii)

The equivalence of the statements below follows from elementary properties of Boolean analysis.

    (a) $(\exists g)$ $(\forall u, V)$ $[f(u, V) = 1 \Longrightarrow u = g(V)]$

    (b) $(\exists g)$ $(\forall u, V)$ $[f(u, V) \leq u' \oplus g(V)]$

    (c) $(\exists g)$ $(\forall V)$ $\begin{bmatrix} f(0, V) & \leq & g'(V) \\ f(1, V) & \leq & g(V) \end{bmatrix}$

    (d) $(\exists g)$ $(\forall V)$ $[f(1, V) \leq g(V) \leq f'(0, V)]$

    (e) $(\forall V)$ $[f(1, V) \leq f'(0, V)]$

    (f) $(\forall V)$ $[f'(1, V) + f'(0, V) = 1]$ *i.e.,* $EDIS(f', \{u\}) = 1$

    (g) $(\forall V)$ $[f(1, V)f(0, V) = 0]$ *i.e.,* $ECON(f, \{u\}) = 0$

(iii) $\Longleftrightarrow$ (iv)

The equivalence of (iii) and (iv) follows directly from Lemma 5.8.1.

$\square$

Statement (d) in the foregoing proof, together with Proposition 3.14.2, leads to

Corollary 7.1.1 *The Boolean equation $u = g(V)$ is a functional consequent of (7.1) if and only if g lies in the interval*

$$f/u \ \leq \ g \ \leq \ (f/u')' \, . \tag{7.4}$$

The function g is thus "incompletely specified" in the terminology of switching theory. The set of g-functions defined by (7.4) may be displayed on a Karnaugh map whose arguments are those appearing explicitly in f/u. A "1" is marked in the cells corresponding to the function f/u, a "0" in the cells corresponding to the function f/u', and an "×" (don't-care) in the remaining cells.

Example 7.1.1 The operation of the most significant stage of a binary two's-complement adder is defined by the system

$$
\begin{aligned}
d &= ab + ac + bc \\
s &= a \oplus b \oplus c \\
u &= abs' + a'b's
\end{aligned}
\tag{7.5}
$$

where a and b are the sign-bits of the addends, s is the sign-bit of the sum, c and d are the input and output carries, respectively, and u is the two's-complement overflow signal. (Non-standard variable-names have been chosen to avoid subscripts.) System (7.5) is equivalent to an equation of the form $f = 1$, where f is given by

$$
\begin{aligned}
f &= a'b'c'd's'u' + a'b'cd'su + a'bc'd'su' + a'bcds'u' + \\
  &\quad ab'c'd'su + ab'cds'u' + abc'ds'u + abcdsu' \,.
\end{aligned}
\tag{7.6}
$$

Formula (7.6) is in Blake canonical form; hence, we may apply criterion (iv) of Theorem 7.1.1 to determine the deducible variables. Each of the variables a, b, c, d, s, u appears in each of the terms of $BCF(f)$; hence, each variable is functionally deducible from the equation $f(a, b, c, d, s, u) = 1$. Let us investigate in particular the possible functions g such that the overflow-variable, u, is given by

$$
u = g(a, b, c, d, s) \,.
$$

Applying Corollary 7.1.1, the set of g-functions is defined by the interval

$$
f/u \le g \le (f/u')' \,,
\tag{7.7}
$$

where f/u and f/u' are given as follows:

$$
\begin{aligned}
f/u &= a'b'cd's + abc'ds' \\
f/u' &= a'b'c'd's' + a'bc'd's + a'bcds' + \\
       &\quad ab'c'd's + ab'cds' + abcds \,.
\end{aligned}
\tag{7.8}
$$

There are 2^{24} 5-variable g-functions in the interval (7.7), one of which, necessarily, is specified by the third equation in (7.5). Another,

$$
u = c \oplus d \,,
\tag{7.9}
$$

is found by inspection of a 5-variable Karnaugh map, using the mapping rules cited earlier. Formula (7.9) is commonly employed in the design of arithmetic circuits. □

Example 7.1.2 (Bennett & Baylis [7], p. 218) What may one infer as to the structure of $A + BC$ from the simultaneous relations $A \leq B + C$, $B \leq A + C$, and $C \leq A + B$?

Solution: We introduce a variable, X, to stand for $A + BC$. Thus the given information is represented by the system

$$
\begin{aligned}
X &= A + BC \\
A &\leq B + C \\
B &\leq A + C \\
C &\leq A + B \, ,
\end{aligned}
$$

which reduces to $f = 1$, where f is given in Blake canonical form by

$$ BCF(f) = A'B'C'X' + BCX + ACX + ABX \, . $$

The variable X, and no other, appears in every term of the foregoing formula. Hence, by criterion (iv) of Theorem 7.1.1, the variable X, and no other, is functionally deducible. Applying Corollary 7.1.1, the set of values of X is specified by the interval

$$ AB + AC + BC \ \leq \ X \ \leq \ A + B + C \, . $$

□

Example 7.1.3 A *sequential* digital circuit is one possessing memory; the present output of such a circuit depends not only on the present input but on the sequence of prior inputs. An *asynchronous* sequential circuit is a sequential circuit whose operation is not timed by an external clock-signal. Changes in the output and memory-state of an asynchronous circuit occur only in response to changes in the input. The memory of an asynchronous circuit is embodied in the values of one or more signal-lines, called *state-variables*, connected in closed feedback-loops. These variables "lock up" in stable states in the intervals between input-changes.

Suppose that an engineer wishes to design an asynchronous circuit having inputs a_1 and a_0 and output z. The output is to have the value 1 if and only if the present value of the pair $a_1 a_0$, viewed as a binary number, is greater than the preceding value; otherwise the output is to be 0. It is assumed that a_1 and a_0 cannot change simultaneously.

Using standard techniques of asynchronous design, the engineer finds that the circuit should realize the equations $z = a_1 a_0 + a_1 y' + a_0 y'$ and $y = a_1 a_0 + a_1 y + a_0 y$, where y is a state-variable. (Note the feedback of the state-variable y implied in the latter equation.) The engineer notes that if a_1, a_0 and y are connected to the 3 inputs of a full adder (a standard component), one of the two outputs of the adder produces the desired y-function. The adder's second output, which we label s, produces the function $s = a_1 \oplus a_0 \oplus y$. The engineer decides to use the full adder to generate the y function; the second adder-output, s, is "free," so he will use the information it provides, if he can, to assist in generating the output z. He has now accumulated the following set of Boolean equations characterizing the design:

$$
\begin{aligned}
z &= a_1 a_0 + a_1 y' + a_0 y' \\
y &= a_1 a_0 + a_1 y + a_0 y \qquad\qquad (7.10) \\
s &= a_1 \oplus a_0 \oplus y \, .
\end{aligned}
$$

This set of equations is equivalent to the single equation $f = 1$, where f is given, in Blake canonical form, by

$$
\begin{aligned}
BCF(f) \;=\; & a_1' a_0' s' y' z' + a_1 a_0' s y' z + a_1' a_0 s y' z + \\
& a_1' a_0 s' y z' + a_1 a_0' s' y z' + a_1 a_0 s y z \, .
\end{aligned}
$$

The variable z appears in every term of the foregoing formula; by criterion (iv) of Theorem 7.1.1, therefore, z is functionally deducible. Applying Corollary 7.1.1, the set of deducible z-values is the interval $[f/z, \, (f/z')']$, where

$$
\begin{aligned}
f/z &= a_1 a_0' s y' + a_1' a_0 s y' + a_1 a_0 s y \\
(f/z')' &= (a_1' a_0' s' y' + a_1' a_0 s' y + a_1 a_0' s' y)' \\
&= s + a_1 a_0 + a_1 y' + a_0 y' + a_1' a_0' y \, .
\end{aligned}
$$

The foregoing formulas show that one member of $[f/z, \, (f/z')']$ is s; hence,

$$
z = s \, ,
$$

i.e.,

$$a_1 a_0 + a_1 y' + a_0 y' = a_1 \oplus a_0 \oplus y \, ,$$

is deducible from the system (7.10). (The validity of this equality, in view of the constraints imposed by (7.10), should be verified.) Happily for the designer, the adder produces both y and z; the design is complete. $\square$

7.2 Eliminable and Determining Subsets

In this section we consider the following problem: given that a variable u is functionally deducible, what are the sets of variables from which the value of u may be computed? We will call such sets u-*determining subsets*.

7.2.1 u-Eliminable Subsets

Let $\{\{u\}, V, W\}$ be a collection of subsets of $\{x_1, \ldots, x_n\}$, the set of arguments of equation (7.1), having the property that each argument appears in exactly one of the subsets. The subset W may be empty; hence the collection may not be a partition. We say that W is u-*eliminable* from (7.1) provided an equation of the form

$$u = g(V)$$

is a consequent of (7.1). The empty set is trivially u-eliminable from (7.1) if u is functionally deducible from (7.1).

Lemma 7.2.1 *Let $X = \{x_1, \ldots, x_n\}$, $V = \{v_1, \ldots, v_s\}$, $W = \{w_1, \ldots, w_t\}$, and let $\{V, W\}$ be a partition of the argument-set X. Let p and q be Boolean functions and let $p(V, W) = 1$ be consistent. Then the following implications are equivalent for all $V \in \mathbf{B}^s$ and $W \in \mathbf{B}^t$:*

$$p(V, W) = 1 \implies q(V) = 1 \qquad (7.11)$$

$$EDIS(p(V, W), W)) = 1 \implies q(V) = 1 \qquad (7.12)$$

Proof. The following statements are equivalent for all $V \in \mathbf{B}^s$ and $W \in \mathbf{B}^t$:

(a) (7.11)
(b) $p(V, W) \leq q(V)$
(c) $p(V, W) \cdot q'(V) = 0$
(d) $\sum_{A \in \mathbf{B}^t} (p(V, A) \cdot q'(V)) = 0$
(e) $EDIS(p(V, W), W)) \leq q(V)$
(f) (7.12)

The equivalence of (a) and (b), and of (e) and (f), follows from the Extended Verification Theorem (Theorem 5.4.1). The pair (b) and (c) are equivalent by the definition of inclusion, and the pair (c) and (d) because a property true for all values of W is true for any particular value. Finally, the pair (d) and (e) are equivalent by the definition of the disjunctive eliminant. $\square$

Theorem 7.2.1 *The subset W is u-eliminable from equation (7.1) if and only if u is deducible from the equation*

$$EDIS(f, W) = 1 . \qquad (7.13)$$

Proof. W is u-eliminable from (7.1) if and only if the implication

$$f(u, V, W) = 1 \quad \Longrightarrow \quad u = g(V) \qquad (7.14)$$

holds. It follows from Lemma 7.2.1, however, that (7.14) is equivalent to the implication

$$EDIS(f(u, V, W), W) = 1 \quad \Longrightarrow \quad u = g(V) , \qquad (7.15)$$

which establishes the theorem. $\square$

Theorem 7.2.2 *The following statements are equivalent:*

(i)   *W is u-eliminable from (7.1).*
(ii) *$EDIS(ECON(f', W), \{u\}) = 1$.*
(iii) *$ECON(EDIS(f, W), \{u\}) = 0$.*
(iv) *u or u' appears in every term of $BCF(EDIS(f, W))$.*

Proof. Follows directly from Theorems 7.1.1 and 7.2.1. $\square$

Example 7.2.1 System (7.5) in Example 7.1.1 was found to possess the consequent $u = c \oplus d$. Thus a u-eliminable subset is $\{a, b, s\}$. Let us employ criterion (iv) of Theorem 7.2.2 to verify that $\{a, b, s\}$ is u-eliminable:

$$
\begin{aligned}
BCF(EDIS(f, \{a, b, s\})) &= BCF(c'd'u' + cd'u + c'd'u' + cdu' + \\
&\quad\quad c'd'u' + cdu' + c'du + cdu') \\
&= c'd'u' + cd'u + cdu' + c'du .
\end{aligned}
$$

The variable u appears in each term of $BCF(EDIS(f, \{a, b, s\}))$; hence, $\{a, b, s\}$ is u-eliminable. □

Statement (iii) of Theorem 7.2.2, together with Theorem 5.8.6, leads to

Corollary 7.2.1 *The argument-subset W is u-eliminable from (7.1) if and only if the condition*

$$
EDIS(f/u', W) \cdot EDIS(f/u, W) = 0 \tag{7.16}
$$

is satisfied identically.

7.2.2 u-Determining Subsets

Consider as before a collection $\{\{u\}, V, W\}$ of subsets of the set $\{x_1, \ldots, x_n\}$ of arguments of equation (7.1), having the property that each argument appears in exactly one of the subsets. As before the subset W may be empty. We say that V is a *u-determining subset* of the arguments of (7.1) provided W is u-eliminable from (7.1).

The subset W is u-eliminable, by Corollary 7.2.1, if and only if condition (7.16) is satisfied. Let us assume that f/u' and f/u are expressed as sum-of-products formulas. From Theorem 5.8.4, therefore, the disjunctive eliminants $EDIS(f/u', W)$ and $EDIS(f/u, W)$ are formed as sum-of-products formulas from f/u' and f/u, respectively, by deleting all literals corresponding to letters in W (any term all of whose literals are thus deleted is replaced by 1). Then W is u-eliminable if and only if the product of any term of $EDIS(f/u', W)$ and any term of $EDIS(f/u, W)$ is zero. Such a product is zero, however, if and only if at least one letter, call it x, appears opposed in the two terms, *i.e.*, if the literal x appears in one term and the literal x' appears in the other.

The foregoing observations are the basis for a procedure, outlined below, which generates the minimal u-determining subsets. This procedure employs Boolean calculations to arrive at a family of minimal sets; variants of this approach have been used to find irredundant formulas [154], maximal compatibles [129], and maximal independent subgraphs [126, 127, 156, 211].

7.2.3   Calculation of Minimal u-Determining Subsets

We assume that u is functionally deducible from (7.1). The following steps produce a sum-of-products formula, F_u, each of whose terms corresponds to a minimal u-determining subset.

1. Express f/u and f/u' as sum-of-products formulas, viz.,

$$f/u \;=\; \sum_{i=1}^{M} p_i$$

$$f/u' \;=\; \sum_{j=1}^{N} q_j$$

   where $p_1, \ldots, p_M$ and $q_1, \ldots, q_N$ are terms, *i.e.*, products of literals.

2. Associate with each pair (p_i, q_j) a complement-free alterm (sum of literals) s_{ij} defined by

$$s_{ij} = \sum (\text{letters that appear opposed in } p_i \text{ and } q_j) \,.$$

3. Define a Boolean function F_u by the product-of-sums formula

$$F_u = \prod_{i=1}^{M} \prod_{j=1}^{N} s_{ij} \,.$$

4. Multiply out, to form a complement-free sum-of-products formula for F_u, and delete absorbed terms. With each term of the resulting formula associate a set of arguments having the same letters; the resulting sets are the minimal u-determining subsets.

Example 7.2.2 Let us employ the foregoing procedure to find the minimal u-determining subsets for the system (7.5) of Example 7.1.1. We carry out step 1 by labeling the terms shown in (7.8):

$$
\begin{aligned}
p_1 &= a'b'cd's & q_1 &= a'b'c'd's' \\
p_2 &= abc'ds' & q_2 &= a'bc'd's \\
& & q_3 &= a'bcds' \\
& & q_4 &= ab'c'd's \\
& & q_5 &= ab'cds' \\
& & q_6 &= abcds
\end{aligned}
$$

The resulting s_{ij} (step 2) are as follows:

$$
\begin{aligned}
s_{11} &= c+s & s_{21} &= a+b+d \\
s_{12} &= b+c & s_{22} &= a+d+s \\
s_{13} &= b+d+s & s_{23} &= a+c \\
s_{14} &= a+c & s_{24} &= b+d+s \\
s_{15} &= a+d+s & s_{25} &= b+c \\
s_{16} &= a+b+d & s_{26} &= c+s
\end{aligned}
$$

Carrying out step 3, and deleting repeated factors, we obtain

$$ F_u = (c+s)(b+c)(b+d+s)(a+c)(a+d+s)(a+b+d) . $$

The result of multiplying out and deleting absorbed terms is

$$ F_u = abc + cd + abs + bcs + acs , $$

to which correspond the minimal u-determining subsets $\{a,b,c\}$, $\{c,d\}$, $\{a,b,s\}$, $\{b,c,s\}$, and $\{a,c,s\}$. The third subset is the one implied by the third equation of (7.5); the second is the one arrived at via Karnaugh-mapping in Example 7.1.1. $\square$

It has come to the author's attention that the method given above for calculating minimal u-determining subsets was given (in a different context) by Halatsis and Gaitanis [76], who called such subsets *minimal dependence sets*.

Exercises

1. (Halatsis & Gaitanis [76]) A switching circuit has inputs $x_1, \ldots, x_6$ and output z. The behavior of the circuit is specified in terms of an orthogonal pair $\{\phi_0, \phi_1\}$ of 6-variable switching functions as follows:

   - If $\phi_0(x_1, \ldots, x_6) = 1$, then $z = 0$.
   - If $\phi_1(x_1, \ldots, x_6) = 1$, then $z = 1$.
   - Otherwise, $z = 0$ or $z = 1$.

   Given that

   $$
   \begin{aligned}
   \phi_0 &= x_1' x_2' x_3' x_4 x_5 x_6 + x_1' x_2' x_3 x_4 x_5' x_6 + x_1 x_2 x_3' x_4' x_5 x_6' \\
   \phi_1 &= x_1' x_2 x_3' x_4 x_5' x_6' + x_1' x_2 x_3 x_4' x_5' x_6 + x_1' x_2 x_3 x_4' x_5 x_6' + \\
          &\quad x_1 x_2' x_3' x_4 x_5' x_6' + x_1 x_2' x_3 x_4' x_5' x_6 + x_1 x_2' x_3 x_4' x_5 x_6' \ ,
   \end{aligned}
   $$

   show that the minimal z-determining subsets are $\{x_1, x_2\}$, $\{x_3, x_4, x_5\}$, and $\{x_3, x_4, x_6\}$.

2. The clocked D-latch is a digital circuit whose excitation-logic is specified by the equations

   $$
   \begin{aligned}
   R &= CD \\
   S &= CD' \ .
   \end{aligned}
   $$

   C is the clock-input, D is the excitation-input, and S and R ("Set" and "Reset") are output-signals.

   (a) Which of the four variables C, D, R, S is functionally deducible from the remaining three?

   (b) For each of each of the variables identified in (a),

      i. Express the set of deduced values of that variable as an interval involving the remaining variables.

      ii. Find the minimal determining subsets for that variable.

Chapter 8

Boolean Identification

We have been concerned until now with techniques of Boolean reasoning. In this chapter and the next, we apply those techniques to the solution of particular kinds of problems. The problems are chosen to illustrate the techniques; no attempt is made to catalogue the problem-areas to which Boolean reasoning might usefully be applied. In the present chapter, we consider how a Boolean "black box" may be identified by means of an adaptive input-output experiment.

Let us suppose that each system or process (which we shall call a *transducer*) in a certain class is equipped with a collection $X = (x_1, x_2, \ldots, x_m)$ of binary inputs, together with a single binary output, z. Each such transducer is characterized by the relation

$$z = f(x_1, x_2, \ldots, x_m) \tag{8.1}$$

for some Boolean function f. The inputs can be manipulated by an experimenter and the resulting output observed.

We assume that the transducers in the given class are described by a *Boolean model, i.e.,* a Boolean equation

$$\phi(X, Y, z) = 0 \ , \tag{8.2}$$

where Y is a vector $(y_1, y_2, \ldots, y_n)$ of binary parameters. For each parameter-setting, $Y = A \in \{0, 1\}^n$, the model (8.2) implies equation (8.1) for some Boolean function f. The mapping from parameter-setting to transducer is typically not reversible; a given transducer may be described by more than one parameter-setting. The transducers thus partition the parameter-settings into equivalence-classes.

We suppose an unidentified transducer in the given class to be presented to an *experimenter*, who is presumed to have perfect powers of deduction. The experimenter knows the function ϕ in equation (8.2); thus he knows how to characterize the entire class parametrically. He does not, however, know the function f in equation (8.1), which characterizes the transducer in his possession. His task is to *identify* that transducer, *i.e.*, to determine f and the associated equivalence-class of Y-values, by means of an input-output experiment.

This model may be employed in a variety of diagnostic applications. The transducer-class might consist of patients undergoing medical diagnosis, in which case equation (8.2) represents the body of knowledge linking a certain category of diseases with the associated symptoms. Each of of the parameters $y_1, \ldots, y_n$ is in this case a disease, which is either present ($y_i = 1$) or not present ($y_i = 0$) in the patient. Each combination of x-values represents a symptom (or test to elicit that symptom). The output z represents the presence ($z = 1$) or absence ($z = 0$) of the symptom corresponding to the vector X, in the presence of disease-pattern Y. This formulation of the problem of medical diagnosis is a variation of that given by Ledley [116, 117, 119]; Ledley advocated the systematic use of Boolean methods in diagnosis as early as 1954 [113].

The problem of diagnosing multiple stuck-type faults in a combinational logic-circuit was formulated as one of Boolean identification by Poage [157] and by Bossen & Hong [15]. The class of possible circuits (faulty and fault-free) is characterized by a system of equations reducible to the model (8.2); the vector Y consists of parameters each of which indicates the stuck-condition (stuck at 1, stuck at 0, or normal) at a point in the circuit. Breuer, Chang, and Su [22] proposed a method for solving the associated Boolean equations, based on an input-output experiment. Kainec [96] has developed a system (written in the Scheme language) which locates multiple faults in a combinational circuit, using Poage's model and Bossen & Hong's checkpoint-concept, by means of adaptive experiment. Kainec's system accepts a description of a combinational circuit, either by means of Boolean equations or by a VHDL (VHSIC[1] Hardware Description Language) specification; the system then assigns internal check-points, suggests test-inputs to the experimenter, accepts the test-results, and provides a report at the end of the experiment on the nature and location (to within an equivalence-class) of faults in the circuit.

---

[1] *Very High-Speed Integrated Circuit*

The problem of identifying the parameters in a Boolean model for the adrenal gland is discussed by Gann, Schoeffler, and Ostrander [63, 177]. In discussing this model, they note that "two essential ingredients of physiological research are the formulation of hypotheses about the operation of the system and the experimental testing of these hypotheses resulting in either its verification (followed by further testing) or else a change in the hypothesis to account for the latest observations. A problem arises when the amount of data is large, for it becomes difficult to determine whether a hypothesis is consistent with all of the observations. Moreover, there are so many possible experiments to perform, it is not practical to choose the experiments at random—rather it is necessary to select so-called critical experiments, the most instructive experiments possible."

Our object in this chapter is to show how Boolean reasoning may be employed to devise such instructive experiments.

8.1 Parametric and Diagnostic Models

Our experimenter employs (8.2) as a repository of knowledge and as a guide to experimentation, modifying $\phi(X, Y, z)$ as knowledge is acquired. The parameters $y_1, y_2, \ldots, y_n$ in (8.2) may serve only a "curve-fitting" purpose; alternatively, they may represent physical values, *e.g.,* of switch-settings or of binary voltage-levels. The two cases cannot be distinguished by experiment. Let us assume for concreteness that the parameters specify physical properties of the transducer under examination.

The experimenter's tasks are

- to determine the function f relating the output to input, for the existing fixed setting of the parameters, and

- to deduce as much as possible concerning the parameter-settings.

8.1.1 Parametric Models

To be useful for identification, equation (8.2) should imply the equation

$$z = \hat{f}(X, Y) \tag{8.3}$$

for some Boolean function $\hat{f}$; equivalently, the variable z should be functionally deducible (*cf.* Section 7.1) from (8.2). We say that (8.2) is *a parametric model* if such is the case. If (8.2) is a parametric model, then it implies an input-output relation of the form $z = f(X)$ for any fixed setting of the parameter-vector Y.

Theorem 8.1.1 *Equation (8.2) is a parametric model if and only if the condition*

$$EDIS(\phi, \{z\}) = 1 \qquad (8.4)$$

is satisfied identically.

Proof.   Follows directly from Theorem 7.1.1. $\square$

If (8.2) is a parametric model, then the function $\hat{f}$ in the implied relation (8.3) is any member of the interval $[\phi'(X, Y, 1),\ \phi(X, Y, 0)]$ (*cf.* the proof of Theorem 7.1.1).

Example 8.1.1 Let $\phi(x_1, y_1, y_2, z) = 0$ describe a class of Boolean transducers, the function ϕ being given by

$$\phi = y_1' z' + x y_2 z + x y_2' z' + x' y_1 z \ .$$

Substituting $z = 0$ and $z = 1$ in turn, we derive

$$\phi(X, Y, 0) = y_1' + x y_2'$$
$$\phi(X, Y, 1) = x' y_1 + x y_2 \ .$$

The functions $\phi(X, Y, 0)$ and $\phi(X, Y, 1)$ sum to 1 identically; hence equation (8.4) is an identity. By Theorem 8.1.1, therefore, $\phi = 0$ is a parametric model. Applying Corollary 7.1.1, this model implies the relation (8.3) provided $\hat{f}$ is in the range

$$\phi'(X, Y, 1) \leq \hat{f} \leq \phi(X, Y, 0) \ ,$$

i.e.,

$$x' y_1' + x y_2' \leq \hat{f} \leq y_1' + x y_2' \ .$$

8.1.2 The Diagnostic Axiom

The resultant of elimination of z from the Boolean model (8.2) is

$$ECON(\phi(X,Y,z),\{z\}) = 0 . \tag{8.5}$$

This equation expresses the knowledge relating X and Y, in the absence of knowledge concerning z, that is deducible from (8.2). We assume, however, that X is freely manipulable; hence X is unconstrained if nothing is known about the value of z. Equation (8.5) is therefore universally quantified in X, *i.e.*, (8.5) holds for all X in $\{0,1\}^m$. When combined with Theorem 4.8.8, these observations lead to what we shall call the *diagnostic axiom*:

If $\phi(X,Y,z) = 0$ represents a class of transducers, then the parameter-vector Y is constrained by the equation

$$EDIS(ECON(\phi(X,Y,z),\{z\}),X) = 0 . \tag{8.6}$$

8.1.3 Diagnostic Equations and Functions

The diagnostic axiom is based on the "physics" of the identification-problem; a parametric model may not itself constrain Y as announced by (8.6). We call the model (8.2) a *diagnostic equation*, however, (and ϕ a *diagnostic function*), in case it is a parametric model that satisfies the diagnostic axiom, *i.e.*, in case it is a parametric model that implies (8.6). Thus (8.2) is a diagnostic equation if and only if the conditions

$$EDIS(\phi,\{z\}) = 1 \tag{8.7}$$
$$EDIS(ECON(\phi,\{z\}),X) \leq \phi \tag{8.8}$$

are satisfied identically.

Theorem 8.1.2 *The model* (8.2) *is a diagnostic equation if and only if it is a parametric model and there are Boolean functions $\hat{f}$ and $\hat{g}$ for which* (8.2) *is equivalent to the system*

$$z = \hat{f}(X,Y) \tag{8.9}$$
$$0 = \hat{g}(Y) . \tag{8.10}$$

Proof. Equation (8.2) is equivalent to the system (8.9, 8.10) if and only if the equation

$$\phi(X,Y,z) = (z \oplus \hat{f}(X,Y)) + \hat{g}(Y) \tag{8.11}$$

is an identity. Suppose (8.2) is a diagnostic equation, *i.e.*, suppose that conditions (8.7) and (8.8) hold. Then for the choices

$$\hat{f}(X,Y) = \phi(X,Y,0) \tag{8.12}$$
$$\hat{g}(Y) = \phi(X,Y,0) \cdot \phi(X,Y,1) \tag{8.13}$$

the following calculations verify (8.11):

$$
\begin{aligned}
(z \oplus \hat{f}(X,Y)) + \hat{g}(Y) &= (z \oplus \phi(X,Y,0)) + (\phi(X,Y,0) \cdot \phi(X,Y,1)) \\
&= z'(\phi(X,Y,0)) + z(\phi'(X,Y,0) + \phi(X,Y,1)) \\
&= z' \cdot \phi(X,Y,0) + z \cdot \phi(X,Y,1) \\
&= \phi(X,Y,z).
\end{aligned}
$$

To obtain the third line from the line before, we impose condition (8.7), in the form $\phi'(X,Y,0) \cdot \phi'(X,Y,1) = 0$. If (8.2) is a diagnostic equation, therefore, there are Boolean functions $\hat{f}(X,Y)$ and $\hat{g}(Y)$ such that (8.2) is equivalent to the system (8.9, 8.10). To prove the converse, let us set $z = 0$ and $z = 1$ successively in (8.11), to deduce the system

$$
\begin{aligned}
\phi(X,Y,0) &= \hat{f}(X,Y) + \hat{g}(Y) \\
\phi(X,Y,1) &= \hat{f}'(X,Y) + \hat{g}(Y).
\end{aligned}
$$

Hence

$$
\begin{aligned}
EDIS(\phi,\{z\}) &= \phi(X,Y,0) + \phi(X,Y,1) = 1 \\
ECON(\phi,\{z\}) &= \phi(X,Y,0) \cdot \phi(X,Y,1) = \hat{g}(Y).
\end{aligned}
$$

The upper equation verifies condition (8.7). The lower equation shows that $ECON(\phi,\{z\})$ is independent of X; hence

$$EDIS(ECON(\phi,\{z\}),X) = ECON(\phi,\{z\}).$$

By Theorem 4.8.4, therefore, condition (8.8) is verified. $\square$

   It is left as an exercise to show that if (8.2) is diagnostic, then the function $\hat{f}$ in equation (8.9) is any member of the interval

$$\phi'(X,Y,1) \leq \hat{f}(X,Y) \leq \phi(X,Y,0),$$

and that the function $\hat{g}$ in equation (8.10) is uniquely specified by

$$\hat{g}(Y) = ECON(\phi,\{z\}) = \phi(X,Y,0) \cdot \phi(X,Y,1).$$

8.1.4 Augmentation

A parametric model defining a class of transducers may fail to be diagnostic. Any parametric model, $\phi(X, Y, z) = 0$, may be converted, however, to a diagnostic model, $AUG(\phi)(X, Y, z) = 0$, which specifies the same class of transducers. The function $AUG(\phi)$ is defined as follows:

$$AUG(\phi) = \phi + EDIS(ECON(\phi, \{z\}), X). \qquad (8.14)$$

We call $AUG(\phi)$ the *augmentation* of ϕ.

To show that $AUG(\phi) = 0$ is diagnostic, we require a preliminary result:

Lemma 8.1.1 *Let $\phi(X, Y, z)$ be a Boolean function. Then*

$$ECON(AUG(\phi), \{z\}) = EDIS(ECON(\phi, \{z\}), X). \qquad (8.15)$$

Proof. In the following development, denote by $h(Y)$ the function $EDIS(ECON(\phi, \{z\}), X)$:

$$
\begin{aligned}
ECON(AUG(\phi), \{z\}) &= ECON((\phi(X, Y, z) + h(Y)), \{z\}) \\
&= [\phi/z' + h(Y)] \cdot [\phi/z + h(Y)] \\
&= (\phi/z' \cdot \phi/z) + h(Y) \\
&= ECON(\phi, \{z\}) + EDIS(ECON(\phi, \{z\}), X) \\
&= EDIS(ECON(\phi, \{z\}), X).
\end{aligned}
$$

The last line is deduced from the one preceding because $\alpha \leq EDIS(\alpha, X)$ for any Boolean function α (Theorem 4.8.4); the remaining calculations resort only to elementary Boolean algebra. $\square$

Theorem 8.1.3 *Let $\phi(X, Y, z) = 0$ be a parametric model defining a class of transducers. Then $AUG(\phi)(X, Y, z) = 0$ is a diagnostic model defining the same class of transducers.*

Proof. The equation $AUG(\phi)(X, Y, z) = 0$ is equivalent to the system

$$
\begin{aligned}
\phi(X, Y, z) &= 0 \\
EDIS(ECON(\phi(X, Y, z), \{z\}), X) &= 0 \,;
\end{aligned}
$$

hence it follows from the diagnostic axiom that $AUG(\phi)(X, Y, z) = 0$ defines the same class of transducers as does $\phi(X, Y, z) = 0$. To show that

$AUG(\phi)(X, Y, z) = 0$ is a diagnostic model, we show that the following are identities:

$$EDIS(AUG(\phi), \{z\}) \;=\; 1 \qquad\qquad (8.16)$$

$$EDIS(ECON(AUG(\phi), \{z\}), X) \;\leq\; AUG(\phi). \qquad (8.17)$$

It is known that $EDIS(\phi, \{z\}) = 1$, inasmuch as $\phi(X, Y, z) = 0$ is a parametric model. Hence

$EDIS(AUG(\phi), \{z\})$

$$\begin{aligned}
&= \; EDIS(\phi + EDIS(ECON(\phi, \{z\}), X), \{z\}) \\
&= \; EDIS(\phi, \{z\}) + EDIS(EDIS(ECON(\phi, \{z\}), X), \{z\}) \\
&= \; 1 + EDIS(EDIS(ECON(\phi, \{z\}), X), \{z\}) \\
&= \; 1,
\end{aligned}$$

verifying (8.16). We begin the verification of (8.17) by expanding its left member, invoking Lemma 8.1.1 and the identity $EDIS(EDIS(\alpha, X), X) = EDIS(\alpha, X)$:

$$\begin{aligned}
EDIS(ECON(AUG(\phi), \{z\}), X) \;&=\; EDIS(EDIS(ECON(\phi, \{z\}), X), X) \\
&=\; EDIS(ECON(\phi, \{z\}), X).
\end{aligned}$$

We make use of definition (8.14) to complete the expansion of (8.17) as follows:

$$EDIS(ECON(\phi, \{z\}), X) \leq \phi + EDIS(ECON(\phi, \{z\}), X).$$

This formula is clearly an identity; hence $AUG(\phi)(X, Y, z) = 0$ is diagnostic, completing the proof. $\square$

Example 8.1.2 Let us consider again the parametric model discussed in Example 8.1.1. The function $EDIS(ECON(\phi(X, Y, z), \{z\}), X)$ evaluates in this case to $y_1' y_2$, which is not included in ϕ, *i.e.*, the condition (8.8) is not satisfied. Thus ϕ is not diagnostic. The augmentation of ϕ is

$$\begin{aligned}
AUG(\phi) \;&=\; \phi + EDIS(ECON(\phi, \{z\}), X) \\
&=\; (y_1' z' + x y_2 z + x y_2' z' + x' y_1 z) + (y_1' y_2) \\
&=\; y_1' z' + y_2 z + x y_2' z' + x' y_1 z.
\end{aligned}$$

The model $AUG(\phi) = 0$ is diagnostic; hence, it is equivalent to the system (8.16, 8.17), where $\hat{f}$ and $\hat{g}$ are specified as follows:

$$\begin{aligned}
x y_2' + y_1' y_2 \;&\leq\; \hat{f} \;\leq\; x y_2' + y_1' \\
\hat{g} \;&=\; y_1 y_2'.
\end{aligned}$$

8.2 Adaptive Identification

We assume that the experimenter supplies a sequence $(A_1, A_2, \ldots)$ of X-values to the transducer under test and observes the corresponding sequence $(f(A_1), f(A_2), \ldots)$ of z-values. We call the X-values *test-inputs* and we call the pair $(A_i, f(A_i))$ a *test*. A sequence of tests for which the test-inputs are distinct will be called an *experiment*.

A given test $(A_i, f(A_i))$ may supply no new information; $f(A_i)$ may be deducible, that is, from tests earlier in the experiment and from information supplied prior to the experiment. It is clear that there are situations, e.g., medical diagnosis, in which such superfluous tests should be avoided. In this section we describe a procedure for conducting an *adaptive* experiment, *i.e.*, one in which the selection of each test-input is based on the outcomes of earlier tests. Each test in the resulting experiment is guaranteed to supply information not deducible from earlier tests or from information supplied prior to testing.

We call an experiment *definitive* if, upon its completion, the experimenter has enough information to deduce the Boolean function f, *i.e.*, if he can specify $f(A_i)$ for any test-input A_i in $\{0, 1\}^m$. An exhaustive experiment, *i.e.*, one which includes all 2^m test-inputs in $\{0, 1\}^m$, is clearly definitive. Our object, however, is to construct definitive experiments based on a subset (preferably a small subset) of all possible inputs; those test-outcomes not found by experiment can be found if needed by deduction.

8.2.1 Initial and Terminal Specifications

After test i, the experimenter's knowledge concerning the transducer is represented by a diagnostic model, *viz.*,

$$\phi_i(X, Y, z) = 0 . \tag{8.18}$$

The index i indicates the number of tests already conducted in the experiment. The transducer is thus characterized initially by the model $\phi_0(X, Y, z) = 0$. Because ϕ_0 is diagnostic, there are Boolean functions $\hat{f}_0$ and $\hat{g}_0$ such that the initial model is equivalent to the system

$$z = \hat{f}_0(X, Y) \tag{8.19}$$
$$0 = \hat{g}_0(Y) . \tag{8.20}$$

We call the pair (8.19, 8.20) an *initial specification* of the system, and we refer to $\hat{f}_0$ and $\hat{g}_0$ as *initial functions*.

Equations (8.19, 8.20) express the information available at the beginning of an experiment. Equation (8.20) specifies the y-constraints, and (8.19) the dependence of z upon X and Y, known prior to testing. The function $\hat{f}_0$ is not unique, but all allowable choices of $\hat{f}_0$ induce the same dependence of z upon X and Y for values of Y satisfying (8.20). The constraint (8.20) introduces "don't-care" conditions, in other words, into the specification of $\hat{f}_0$ (*cf.* Example 8.1.2).

Theorem 8.2.1 *Let a class of transducers be described in terms of initial functions $\hat{f}_0$ and $\hat{g}_0$. At the completion of a definitive experiment, the information concerning the transducer under test is expressed by the system*

$$z = f(X) \tag{8.21}$$
$$0 = g(Y), \tag{8.22}$$

where f and g are unique Boolean functions and where g is given in terms of $\hat{f}_0$, $\hat{g}_0$, and f by the relation

$$g(Y) = \hat{g}_0(Y) + EDIS((f(X) \oplus \hat{f}_0(X,Y)), X). \tag{8.23}$$

Proof. The information acquired by means of any definitive experiment is the same as that acquired by means of an exhaustive experiment; the latter information is expressed by the system

$$
\begin{aligned}
z &= \hat{f}_0(X,Y) \\
0 &= \hat{g}_0(Y) \\
f(0,\dots,0,0) &= \hat{f}_0(0,\dots,0,0,Y) \\
f(0,\dots,0,1) &= \hat{f}_0(0,\dots,0,1,Y) \\
f(0,\dots,1,0) &= \hat{f}_0(0,\dots,1,0,Y) \\
&\;\;\vdots \\
f(1,\dots,1,1) &= \hat{f}_0(1,\dots,1,1,Y).
\end{aligned}
\tag{8.24}
$$

The first two equations are the initial specifications. Each of the 2^m remaining equations denotes the information supplied by a single test. In view of the latter equations, and making use of minterm expansion (*cf.* Theorem 2.9.1), the first equation may be re-cast equivalently as follows:

$$z = \hat{f}_0(X,Y)$$
$$= \sum_{A\in\{0,1\}^m} \hat{f}_0(A,Y)X^A$$
$$= \sum_{A\in\{0,1\}^m} f(A)X^A$$
$$= f(X) .$$

By Theorem 4.3.1, the remaining $2^m + 1$ equations in system (8.24) are equivalent to the single equation

$$\hat{g}_0(Y) + \sum_{A\in\{0,1\}^m} (f(A) \oplus \hat{f}_0(A,Y)) = 0 ,$$

which is re-written equivalently, in view of Theorem 4.8.1, as

$$\hat{g}_0(Y) + EDIS((f(X) \oplus \hat{f}_0(X,Y)), X) = 0 . \tag{8.25}$$

The system (8.24) is thus equivalent to the system (8.21, 8.22), with $g(Y)$ expressed by the left side of (8.25). The uniqueness of f and g is guaranteed by the disjointness of the arguments appearing in (8.21) from those in (8.22). □

We call the pair (8.21, 8.22) the *terminal specification* of the system, and we refer to f and g as the *terminal functions*. Equation (8.21) expresses the dependence of z upon X, for the existing fixed value of Y; equation (8.22) specifies the parameter-vector Y to within an equivalence-class, namely, the set of solutions of $0 = g(Y)$.

Example 8.2.1 Let $\hat{f}_0$, $\hat{g}_0$, and f be given by the formulas

$$\hat{f}_0(X,Y) = x_1 y_1' + x_2' y_2 + x_1' x_2' y_3$$
$$\hat{g}_0(Y) = y_1 y_3$$
$$f(X) = x_1 + x_2' .$$

Applying Theorem 8.2.1,

$$g(Y) = y_1 y_3 + EDIS([(x_1 + x_2') \oplus (x_1 y_1' + x_2' y_2 + x_1' x_2' y_3)], X)$$
$$= y_1 y_3 + EDIS((x_1' x_2' y_2' y_3' + x_1 x_2 y_1 + x_1 y_1 y_2'), X)$$
$$= y_1 y_3 + (y_2' y_3' + y_1 + y_1 y_2')$$
$$= y_1 + y_2' y_3' .$$

8.2.2   Updating the Model

The experimenter's knowledge after the test $(A_i, f(A_i))$ is expressed by the pair of equations

$$\phi_{i-1}(X, Y, z) = 0 \tag{8.26}$$

$$\phi_{i-1}(A_i, Y, f(A_i)) = 0 . \tag{8.27}$$

The first equation expresses the experimenter's knowledge prior to the test; the second expresses the new information supplied by the test. The new state of affairs is therefore expressed by the model

$$\phi_i(X, Y, z) = 0 , \tag{8.28}$$

where ϕ_i, the updated model-function, is given by the recursion

$$\phi_i(X, Y, z) = \phi_{i-1}(X, Y, z) + \phi_{i-1}(A_i, Y, f(A_i)) . \tag{8.29}$$

If $\phi_{i-1} = 0$ is diagnostic and ϕ_i is defined by (8.29), then $\phi_i = 0$ is also diagnostic (the proof is left as an exercise).

Example 8.2.2 Assume again the initial functions $\hat{f}_0$ and $\hat{g}_0$, and the terminal function f, given in Example 8.2.1. The experimenter knows $\hat{f}_0$ and $\hat{g}_0$ (or, equivalently, the initial model-function, ϕ_0), but does not know f; his object is to determine the terminal functions f and g. The initial model-function, ϕ_0, is given by

$$
\begin{aligned}
\phi_0 &= (z \oplus \hat{f}_0) + \hat{g}_0 \\
&= x_2' y_2 z' + x_1' x_2 z + x_2' y_3 z' + x_2 y_1 z + y_1 y_2' z + x_1 y_1' z' + x_1' y_2' y_3' z + y_1 y_3 .
\end{aligned}
$$

The experimenter applies the test-vector $X = (0, 0)$, chosen at random, and observes that the resulting output-value is $z = 1$. The information supplied by this test is expressed by equation (8.27), which in the present instance takes the form

$$y_1 y_3 + y_2' y_3' = 0 . \tag{8.30}$$

Applying the recursion (8.29) enables the experimenter to calculate ϕ_1:

$$
\begin{aligned}
\phi_1 &= \phi_0 + y_1 y_3 + y_2' y_3' \\
&= x_2' z' + x_2 y_1 z + x_1' x_2 z + x_1 y_1' z' + y_1 y_3 + y_2' y_3' .
\end{aligned}
$$

8.2.3 Effective Inputs

The updated model $\phi_1(X, Y, z) = 0$ in Example 8.2.2, resulting from the test $((0,0),1)$ (*i.e.*, $X = (0,0) \implies z = 1$), supplies strictly more information concerning the transducer than does the initial model, $\phi_0(X, Y, z) = 0$. If the test-input had been chosen to be $X = (0,1)$, however, the test would have supplied no new information; the output ($z = 0$) is predictable from the initial model. The updated model-function resulting from the test $((0,1),0)$, is thus the same as the initial model-function, *i.e.*,

$$\phi_0 + y_1 y_3 = \phi_0 .$$

We wish to avoid such useless tests; hence we desire that the test-vector $X = A_{i+1}$ be chosen so that the resulting output is not predictable from knowledge of the current model, $\phi_i = 0$. Such a test-vector will be called *effective*.

Input-equation. To assess the effectiveness of test-inputs, we associate a Boolean *input-equation*,

$$\psi_i(X) = 0 , \tag{8.31}$$

with the model $\phi_i = 0$. The *input-function, ψ_i,* is defined as follows:

$$\psi_i(X) = EDIS(ECON(\phi_i, Y), \{z\}) . \tag{8.32}$$

The utility of the input-equation is indicated in the next theorem.

Theorem 8.2.2 *Let the current state of knowledge concerning a class of transducers be expressed by the diagnostic model $\phi_i(X, Y, z) = 0$, and let the Boolean function ψ be defined by (8.32). Then the effective inputs are the solutions of the equation*

$$\psi_i(X) = 0 . \tag{8.33}$$

Proof. The relation between X and z is found by eliminating Y from the current model, $\phi_i = 0$. The resultant is

$$ECON(\phi_i, Y) = 0 . \tag{8.34}$$

The input-vector $X = A_{i+1}$ supplied for the next test is effective in case the output z is not functionally deducible from (8.34). By Theorem 8.1.1, z is functionally deducible from (8.34) if and only if the condition

$$EDIS(ECON(\phi_i, Y), \{z\}) = 1 \tag{8.35}$$

is satisfied. The left side of (8.34), *i.e.*, $\psi_i(X)$, is a two-valued function; hence the denial of (8.34) (the necessary and sufficient condition for an effective input) is expressed by (8.33). $\square$

Example 8.2.3 Applying (8.32), the input-function ψ_0 corresponding to the initial model given in Example 8.2.2 is

$$\psi_0(X) = x_1' x_2 \, .$$

Thus the effective test-inputs, *i.e.*, the solutions of $x_1' x_2 = 0$, are $(0,0), (1,0)$, and $(1,1)$. The test-input employed in Example 8.2.2, $X = (0,0)$, is thus verified as effective. The information supplied by test $((0,0),1)$ is employed in Example 8.2.2 to derive an updated model, $\phi_1 = 0$. The associated input-equation is $\psi_1(X) = 0$, where

$$\psi_1(X) = x_1' + x_2' \, .$$

Thus there is now just one effective test-input, *viz.*, $X = (1,1)$. The experimenter determines that the response to that input is $z = 1$; hence the updated model is $\phi_2 = 0$, where

$$
\begin{aligned}
\phi_2 &= \phi_1(X,Y,z) + \phi_1(1,1,y_1,y_2,y_3,1) \\
&= (x_2' z' + x_2 y_1 z + x_1' x_2 z + x_1 y_1' z' + y_1 y_3 + y_2' y_3') + (y_1 + y_2' y_3') \\
&= x_1 z' + x_2' z' + x_1' x_2 z + y_1 + y_2' y_3' \, .
\end{aligned}
$$

The input-equation is now $\psi_2 = 0$, where ψ_2 is determined by (8.32) to have the value 1. The equation $1 = 0$ has no solutions; thus there are at this point no effective inputs. The experiment $(((0,0),1), ((1,1),1))$ is therefore definitive. The final model-function has the expansion

$$\phi_2 = (z \oplus (x_1 + x_2')) + y_1 + y_2' y_3' \, ;$$

hence the terminal functions are

$$
\begin{aligned}
f(X) &= x_1 + x_2' \\
g(Y) &= y_1 + y_2' y_3' \, .
\end{aligned}
$$

The function f (unknown to the experimenter at the outset) agrees with that given in Example 8.2.1. The equivalence-class of parameter-vectors is the set of solutions of $g(Y) = 0$; for the transducer just identified, the class is $\{(0,0,1),(0,1,0),(0,1,1)\}$.

Terminal model. The number of distinct test-inputs is finite and no effective test-input repeats an earlier one. Thus if all test-inputs are effective, there is some index, $k \leq 2^m$ (as shown in the foregoing example), such that $\psi_k = 1$ is an identity. This indicates that no further effective test-inputs exist, *i.e.*, that the k tests performed thus far constitute a definitive experiment. Consequently we call $\phi_k(X, Y, z) = 0$ a *terminal model*.

Theorem 8.2.3 *Let* $\phi_k(X, Y, z) = 0$ *be a terminal model for a transducer under test. Then the terminal functions* f *and* g *are given by*

$$f(X) = ECON(\phi_k(X, Y, 0), Y) \qquad (8.36)$$
$$g(Y) = ECON(\phi_k(X, Y, z), \{z\}) . \qquad (8.37)$$

Proof. By Theorem 8.2.1, the model $\phi_k(X, Y, z) = 0$ is equivalent to the system (8.21, 8.22), from which we conclude that

$$\phi_k(X, Y, z) = (z \oplus f(X)) + g(Y) .$$

Thus

$$
\begin{aligned}
ECON(\phi_k, Y) &= \prod_{A \in \{0,1\}^n} ((z \oplus f(X)) + g(A)) \\
&= (z \oplus f(X)) + \prod_{A \in \{0,1\}^n} g(A) \\
&= z \oplus f(X) .
\end{aligned}
$$

The last line is derived from the assumption that the equivalence-class of parameter-vectors, *i.e.*, the set of solutions of (8.21), is not empty. Thus the consistency-condition,

$$\prod_{A \in \{0,1\}^n} g(A) = 0 ,$$

(*cf.* Theorem 6.1.1) is satisfied. Hence,

$$
\begin{aligned}
ECON(\phi_k(X, Y, 0), Y) &= 0 \oplus f(X) \\
&= f(X) ,
\end{aligned}
$$

verifying (8.36). The following computations verify (8.37):

$$
\begin{aligned}
ECON(\phi_k, \{z\}) &= ((0 \oplus f(X)) + g(Y)) \cdot ((1 \oplus f(X)) + g(Y)) \\
&= (f(X) + g(Y)) \cdot (f'(X) + g(Y)) \\
&= g(Y) .
\end{aligned}
$$

$\square$

8.2.4   Test-Procedure

Given the initial model, $\phi_0(X, Y, z) = 0$, the experimenter's object is to determine the terminal functions $f(X)$ and $g(Y)$ by carrying out a sequence of effective tests.

The algorithm shown in Figure 8.1 enables the experimenter to choose an effective test-input at each stage of an experiment. The algorithm is based on a diagnostic model-function, $\phi(X, Y, z)$, from which the input-function $\psi(X)$ is derived (subscripts are omitted). The test-input at each stage is obtained by solving the input-equation, $\psi(X) = 0$; the resulting observed output is used to re-calculate $\phi(X, Y, z)$ and $\psi(X)$, after which the process is repeated. The test-inputs derived in this way are necessarily distinct; hence, the updated model becomes equivalent ultimately to a terminal description; this condition is signaled when $\psi(X)$ becomes equal identically to 1.

```
 1      ψ(X) := EDIS(ECON(φ,Y), {z})
 2          while ψ(X) ≠ 1 do
 3              begin
 4                  Select a solution, A, of ψ(X) = 0.
 5                  Apply X = A as a test-input.
 6                  Observe the corresponding output, z = f(A).
 7                  φ(X,Y,z) := φ(X,Y,z) + φ(A,Y,f(A))
 8                  ψ(X) := EDIS(ECON(φ,Y), {z})
 9              end
10          end
11          Return the terminal functions
12                      f(X) = ECON(φ(X,Y,0),Y)
13                      g(Y) = ECON(φ(X,Y,z), {z})
14      end
```

Figure 8.1: Algorithm for a definitive experiment.

Example 8.2.4 Let us apply the foregoing procedure to identify a transducer in the class characterized by the diagnostic equation $\phi_0 = 0$, where ϕ_0 is given by

$$
\begin{aligned}
\phi_0 \; = \; & y_2 y_3' + x_3' y_2 z + x_1' y_3' z + x_1 x_3' y_2' z' + x_1' x_2 y_2' z + x_1 x_3 y_2' z + \\
& x_1 x_2' x_3 z + x_1 x_3 y_1 z + x_1' x_3 y_2 z' + x_2 x_3 y_1' y_2 z' + x_1' x_2' y_2' y_3 z' \; .
\end{aligned}
$$

The following steps (shown with arguments indexed) constitute a definitive experiment:

$$
\begin{aligned}
\psi_0 &= x_1' x_2 x_3' + x_1 x_2' x_3 \\
A_1 &= (1,1,1) \\
f(A_1) &= 0 \\
\phi_1 &= \phi_0 + y_2 y_3' + y_1' y_2 \\
&= y_2 y_3' + y_1' y_2 + x_1 x_3 z + x_3' y_2 z + x_1' y_3' z \\
&\quad + x_1 x_3' y_2' z' + x_1' x_2 y_2' z + x_1' x_3 y_2 z' + x_1' x_2' y_2' y_3 z' \\
\psi_1 &= x_1' x_2 x_3' + x_1 x_3 \\
A_2 &= (0,0,0) \\
f(A_2) &= 1 \\
\phi_2 &= \phi_1 + y_2 + y_3' \\
&= y_2 + y_3' + x_1 x_3 z + x_1 x_3' z' + x_1' x_2 z + x_1' x_2' z' \\
\psi_2 &= 1 \\
f(X) &= ECON(\phi_2(X,Y,0),Y) = x_1 x_3' + x_1' x_2 \\
g(Y) &= ECON(\phi_2(X,Y,z),\{z\}) = y_2 + y_3'
\end{aligned}
$$

The test-vectors A_1 and A_2 were selected arbitrarily from among the solutions, respectively, of $\psi_0 = 0$ and $\psi_1 = 0$. The adaptive nature of this process enables us to identify the transducer after only two tests; an exhaustive experiment would have required $2^3 = 8$ tests.

Exercises

1. Show that if (8.2) is diagnostic, then the function $\hat{f}$ in equation (8.9) is any member of the interval

$$
\phi'(X,Y,1) \le \hat{f}(X,Y) \le \phi(X,Y,0),
$$

and that the function $\hat{g}$ in equation (8.10) is uniquely specified by

$$
\hat{g}(Y) = ECON(\phi,\{z\}).
$$

2. Given that $\phi_{i-1}(X,Y,z) = 0$ is a diagnostic equation and that the Boolean function ϕ_i is defined by

$$
\phi_i(X,Y,z) = \phi_{i-1}(X,Y,z) + \phi_{i-1}(A_i,Y,f(A_i)),
$$

show that $\phi_i = 0$ is also diagnostic.

3. Line 7 of the algorithm shown in Figure 8.1 is the assignment

$$\phi(X,Y,z) := \phi(X,Y,z) + \phi(A,Y,f(A)) .$$

Show that if ϕ is diagnostic, then

$$\phi(X,Y,z) + \phi(A,Y,f(A)) = AUG(\phi(X,Y,z) + X^A z^{f'(A)}) ,$$

whence line 7 of the algorithm may be replaced by

$$\phi(X,Y,z) := AUG(\phi(X,Y,z) + X^A z^{f'(A)}) .$$

Chapter 9

Recursive Realizations of Combinational Circuits

In this chapter we illustrate some applications of Boolean reasoning in the design of multiple-output switching circuits. The stimulus applied to the circuit shown in Figure 9.1 is an input-vector, $X = (x_1, x_2, \ldots, x_m)$, of binary signals; its response is an output-vector, $Z = (z_1, z_2, \ldots, z_n)$, of binary signals. We assume the circuit to be *combinational*, by which we mean that the value of Z at any time is a function of the value of X at that time. A *sequential* circuit, on the other hand, is one for which the value of Z may depend on past values of X as well as on its present value.

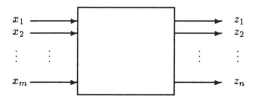

Figure 9.1: Multiple-Output Circuit.

Boolean methods for combinational circuit-design have been investigated for more than fifty years [145, 146, 183, 185]. The field nevertheless remains an area of active research; see Brand [16] and Brayton, *et al.* [18, 20] for summaries of recent work.

It is not our intent to discuss the extensive and highly-developed theory of digital circuit design. The presentation in this chapter is arranged instead with the following objectives in mind:

- to formulate the problem of multiple-output combinational circuit-design within the framework of Boolean reasoning; and

- to show how this formulation may be applied to solve a particular problem in circuit-design for which customary Boolean methods have proved inadequate.

9.1   The Design-Process

Combinational circuit-design customarily proceeds by means of the following steps:

1. **Specification.**   Describe a class of circuits suitable to one's purpose by stating a relation between X and Z. A common specification-format is a truth-table.

2. **Solution.**   Solve the specification, *i.e.,* find a vector switching function, $F = (f_1, f_2, \ldots, f_n)$, such that the system

$$Z = F(X) \tag{9.1}$$

   implies the specification. The solution is chosen so that $f_1, f_2, \ldots, f_n$ can be represented by simplified formulas.

3. **Transformation.**   Transform the solution (9.1) into a system

$$\begin{aligned} Z &= G(X, Y) \\ Y &= H(X, Y) \end{aligned} \tag{9.2}$$

   such that the resultant of elimination of Y from (9.2) (*cf.* Chapter 4) is equivalent to the solution (9.1). The transformation from (9.1) to (9.2) is carried out so as to optimize selected measures of cost and performance. The transformation-process, whether automated or manual, is typically carried out by a sequence of local changes. The elements of the vector $Y = (y_1, y_2, \ldots, y_k)$ represent "internal" signals, introduced as the transformation proceeds.

The problem we consider is how best to use a circuit's output signals—in addition to its inputs—to assist in the generation of its outputs. The advantageous use of available signals—including outputs—is a part of skilled manual design, but has thus far been automated only in limited ways. The text *Synthesis of Electronic Computing and Control Circuits,* published in 1951 by the Staff of the Harvard Computation Laboratory [82] devotes a chapter to this problem; further work has been reported by Ho [85], Kobrinsky [105], Mithani [139], and Pratt [160].

The process of design outlined in this chapter differs in the following respects from the three-step procedure outlined earlier:

- **Specification-Format.** The specification may be expressed initially in a number of ways, *e.g.,* by an enumeration of (X, Z)-pairs, by a truth-table, or by a system of Boolean equations. It is reduced, however, to a single Boolean equation of the form

$$\phi(X, Z) = 1 . \tag{9.3}$$

  We call (9.3) the *normal form* for the specification. Reducing the specification to a single equation enables global dependencies and "don't-care" conditions to be handled uniformly and systematically.

- **Nature of the Transformed System.** The transformed system has the form
$$Z = F(X, Z) \tag{9.4}$$

  rather than that of (9.2). Thus the output-values may depend not only on the values of inputs but also on those of outputs. The function-vector F in (9.4) is chosen to have the following properties:

  - **Stability.** The feedback implied by the presence of Z as an argument of F does not cause oscillation.

  - **Economy.** The total cost of the formulas expressing the functions $f_1, f_2, \ldots, f_k$ is as small as design-constraints will permit.

9.2 Specifications

Let $\mathcal{I} = \{0, 1\}^m$ and $\mathcal{O} = \{0, 1\}^n$ be the input-space and output-space, respectively, of the combinational circuit shown in Fig. 9.1. A *specification* for the circuit is a relation $\mathcal{R}$ from $\mathcal{I}$ to $\mathcal{O}$, *i.e.,* a subset of $\mathcal{I} \times \mathcal{O}$ (*cf.* Section

1.6). Thus an input-output pair (X, Z) is allowed by the specification if and only if $(X, Z) \in \mathcal{R}$

Example 9.2.1 A 2-input AND-gate, for which $X = (x_1, x_2)$ and $Z = z_1$, is specified by the subset

$$\mathcal{R} = \{((0,0),0), ((0,1),0), ((1,0),0), ((1,1),1)\} \qquad (9.5)$$

of $\{0,1\}^2 \times \{0,1\}$. $\square$

A specification $\mathcal{R}$ will be called *complete* in case it defines a function from $\mathcal{I}$ into $\mathcal{O}$, *i.e.*, in case each member of $\mathcal{I}$ appears exactly once as a left element of a pair in $\mathcal{R}$. A specification will be called *incomplete* in case it is not complete, *i.e.*, in case one or both of the following conditions occurs:

(b)   there is a vector $A \in \mathcal{I}$ that appears in more than one pair in $\mathcal{R}$ as a left element; or

(a)   there is a vector $A \in \mathcal{I}$ that fails to appear in any pair in $\mathcal{R}$ as a left element.

These are referred to in circuit-design as "don't-care" conditions. If a vector $A \in \mathcal{I}$ appears more than once as a left member of a pair $(A, Z) \in \mathcal{R}$ (condition (a)), then the circuit may be designed to produce whichever of the corresponding Z-values best meets the designer's objectives. If a vector $A \in \mathcal{I}$ fails to appear as a left element of a pair in $\mathcal{R}$ (condition (b)), then $X = A$ is a forbidden input. The circuit-designer cannot enforce a condition of type (b), which is a constraint on the signal-source producing X.

Any actual m-input, n-output circuit is defined by a set $\mathcal{F} \in 2^{\mathcal{I} \times \mathcal{O}}$ of ordered pairs denoting a function. The circuit will be said to *realize* a specification $\mathcal{R}$ provided the condition

$$(X, Z) \in \mathcal{F} \quad \implies \quad (X, Z) \in \mathcal{R} \qquad (9.6)$$

is satisfied if X appears as a left element of a pair in $\mathcal{R}$.

9.2.1   Specification-Formats

A specification $\mathcal{R}$ may be represented in a number of ways, *e.g.*, by an enumeration of (X, Z)-pairs (as exhibited in Example 9.2.1), by verbal statements, by a predicate-calculus formula, by a truth-table, or by a system of

Boolean equations. The explicit enumeration shown in (9.5), for example, may be represented by the equivalent specification

$$z_1 = x_1 x_2 . \tag{9.7}$$

For most purposes, clearly, the form (9.7) is to be preferred over the enumerative form (9.5)

Example 9.2.2 The information needed to convert between the JK and RST flip-flop types is expressed by the system

$$\begin{aligned} Q'J + QK' &= S + Q'T + QR'T' \\ 0 &= RS + RT + ST \end{aligned} \tag{9.8}$$

of Boolean equations. The first equation relates the next-state behavior of the two flip-flops; the second equation is an excitation-constraint on the RST flip-flop. If $X = (J, K, Q)$ and $Z = (R, S, T)$, then (9.8) specifies the logic to convert from an RST to a JK flip-flop; if $X = (Q, R, S, T)$ and $Z = (J, K)$, then (9.8) specifies the logic to convert from a JK to an RST flip-flop (the directions of these conversions may seem at first glance to be reversed!). $\square$

Normal Form. A specification $\mathcal{R}$, as defined above, is a Boolean constraint (*cf.* Section 4.6). Hence any specification may be expressed as a Boolean equation of the form

$$\phi(X, Z) = 1 . \tag{9.9}$$

The function ϕ is defined in terms of the specification $\mathcal{R}$ as follows: for all $(A, B) \in \{0, 1\}^{m+n}$,

$$\phi(A, B) = 1 \iff (A, B) \in \mathcal{R} . \tag{9.10}$$

The normal-form representation (9.9) is advantageous for a number of reasons. It provides a standardized representation on which to base analysis and synthesis. The function ϕ corresponding to a given specification, $\mathcal{R}$, is unique; the function ϕ, as we shall show, is directly related to a truth-table if expanded in minterms of X. Finally, the normal form provides a uniform and convenient way to represent and deal with "don't-care" conditions.

Henceforth a specification will be assumed to be in normal form if not announced to be otherwise.

Example 9.2.3 The normal form of specification (9.8) is

$$\phi(J, K, Q, R, S, T) = 1 \,, \tag{9.11}$$

where ϕ is given by

$$\phi(J, K, Q, R, S, T) \;=\; \begin{aligned}[t] &J'Q'S'T' + JQ'R'S'T + JQ'R'ST' \\ &+ K'QR'T' + KQRS'T' + KQR'S'T \,. \end{aligned} \tag{9.12}$$

□

Theorem 9.2.1 *Let a specification* $\mathcal{R} \subseteq \mathcal{I} \times \mathcal{O}$ *be represented by the normal form* $\phi(X, Z) = 1$. *Then* ϕ *is given by the expansion*

$$\phi(X, Z) = \sum_{(P,Q) \in \mathcal{R}} X^P Z^Q \,. \tag{9.13}$$

Proof. We show that, for all $(A, B) \in \{0, 1\}^{m+n}$, the equivalence

$$\sum_{(P,Q) \in \mathcal{R}} A^P B^Q = 1 \quad \Longleftrightarrow \quad (A, B) \in \mathcal{R} \tag{9.14}$$

holds. Suppose for $(A, B) \in \{0, 1\}^{m+n}$ that the equation on the left side of (9.14) is satisfied. Each term in the summation has a value of either 0 or 1. We therefore deduce from the relation

$$A^P B^Q \;=\; \begin{cases} 1 & \text{if } A = P \text{ and } B = Q \\ 0 & \text{otherwise} \end{cases} \tag{9.15}$$

that $(A, B) \in \mathcal{R}$. Let us suppose on the other hand that $(A, B) \in \mathcal{R}$. Then (9.15) guarantees that one of the terms in the summation on the left of (9.14), and therefore the summation itself, has the value 1. □

Example 9.2.4 Let a specification be given by the following set of pairs:

$$\mathcal{R} \;=\; \begin{aligned}[t] \{ &((0,0,0),(1,0)), \\ &((0,0,0),(1,1)), \\ &((0,1,0),(1,1)), \\ &((1,0,0),(0,0)), \\ &((1,0,0),(1,0)), \\ &((1,0,1),(0,0)), \\ &((1,0,1),(0,1)), \\ &((1,1,0),(0,1)), \\ &((1,1,1),(0,0)), \\ &((1,1,1),(0,0)), \\ &((1,1,1),(0,0)), \\ &((1,1,1),(0,0)) \} \,. \end{aligned} \tag{9.16}$$

Applying Theorem 9.2.1, $\mathcal{R}$ is equivalent to the normal-form specification $\phi = 1$, where ϕ is given by

$$
\begin{aligned}
\phi \;=\; & x_1' x_2' x_3' z_1 z_2' + x_1' x_2' x_3' z_1 z_2 + x_1' x_2 x_3' z_1 z_2 + x_1 x_2' x_3' z_1' z_2' + \\
& x_1 x_2' x_3' z_1 z_2' + x_1 x_2' x_3 z_1' z_2' + x_1 x_2' x_3 z_1' z_2 + x_1 x_2 x_3' z_1' z_2' + \\
& x_1 x_2 x_3 z_1' z_2' + x_1 x_2 x_3 z_1' z_2 + x_1 x_2 x_3 z_1 z_2' + x_1 x_2 x_3 z_1 z_2 \;.
\end{aligned}
\tag{9.17}
$$

$\square$

Corollary 9.2.1 *A specification (9.9) is complete if and only if the condition*

$$
\phi(A, Z) \text{ is a minterm on the } Z\text{-variables} \qquad (\forall A \in \{0,1\}^m) \tag{9.18}
$$

is satisfied.

Example 9.2.5 Suppose that $X = (J, K, Q)$ and $Z = (R, S, T)$. The specification (9.8) is incomplete because $\phi(0,0,0,R,S,T) = S'T'$ is not a minterm on R, S, T. $\square$

Example 9.2.6 Consider the specification

$$
\begin{aligned}
z_1 &\le x_1' x_2' \\
x_1 + x_2' \;\le\; z_2 &\le x_1 + z_1 \,,
\end{aligned}
\tag{9.19}
$$

for which $X = (x_1, x_2)$ and $Z = (z_1, z_2)$. The normal form for this specification is the equation

$$
\phi(x_1, x_2, z_1, z_2) = 1 \,, \tag{9.20}
$$

in which ϕ is given by

$$
\phi(x_1, x_2, z_1, z_2) = x_1' x_2' z_1 z_2 + x_1' x_2 z_1' z_2' + x_1 z_1 z_2' \,. \tag{9.21}
$$

Thus

$$
\begin{aligned}
\phi(0, 0, z_1, z_2) &= z_1 z_2 \\
\phi(0, 1, z_1, z_2) &= z_1' z_2' \\
\phi(1, 0, z_1, z_2) &= z_1' z_2 \\
\phi(1, 1, z_1, z_2) &= z_1' z_2 \,.
\end{aligned}
\tag{9.22}
$$

Each of the foregoing is a Z-minterm; hence, the specification (9.19) is complete. $\square$

9.2.2 Consistent Specifications

A specification is *consistent* (*cf.* Section 6.1) in case it can be solved for Z in terms of X, *i.e.*, in case there is a system

$$Z = G(X), \tag{9.23}$$

such that the result of substituting (9.23) into (9.9), *i.e.*,

$$\phi(X, G(X)) = 1, \tag{9.24}$$

is an identity. It is a direct consequence of Theorem 6.1.1 that (9.9) is consistent if and only if the condition

$$e(X) = 1 \tag{9.25}$$

is satisfied; the function e is defined by

$$e(X) = EDIS(\phi(X, Z), Z). \tag{9.26}$$

In view of Theorem 9.2.1, equation (9.26) takes the form

$$\begin{aligned} e(X) &= EDIS(\sum_{(P,Q)\in\mathcal{R}} X^P Z^Q, Z) \\ &= \sum_{(P,Q)\in\mathcal{R}} X^P. \end{aligned}$$

Thus specification $\mathcal{R}$ is consistent if and only if the input satisfies the constraint

$$\sum_{(P,Q)\in\mathcal{R}} X^P = 1. \tag{9.27}$$

A specification may be consistent *identically*, *i.e.*, (9.27) may be an identity. It may be also be consistent *physically*, *i.e.*, the input-vector X may be constrained in such a way that (9.27) is satisfied for all values of X that are allowed physically to occur. The question of input-constraints, however, is extrinsic to the specification. We therefore take it as an axiom of circuit-design that a given specification is consistent, *i.e.*, that the input-vector is constrained so as to satisfy condition (9.27).

9.3   Tabular Specifications

To design a circuit realizing a specification $\phi(X, Z) = 1$ entails solving the specification, implicitly or explicitly, for Z in terms of X. Applying the methods discussed in Chapter 6, a general solution of $\phi(X, Z) = 1$, representing the set of all particular solutions, may be expressed by a system of recurrent subsumptions. Because of its recurrent nature (*cf.* Section 6.3) such a general solution is not convenient for locating particular solutions having desired properties. A more convenient form for a general solution is a non-recurrent system

$$
\begin{array}{ccccc}
\alpha_1(X) & \leq & z_1 & \leq & \beta_1(X) \\
\alpha_2(X) & \leq & z_2 & \leq & \beta_2(X) \\
\alpha_3(X) & \leq & z_3 & \leq & \beta_3(X) \\
& & \vdots & & \\
\alpha_n(X) & \leq & z_n & \leq & \beta_n(X) ,
\end{array}
\tag{9.28}
$$

in which the z_i are expressed by independent subsumptions. It is not possible in the general case to express a general solution as a system of the form (9.28). It is always possible, however, to express a general solution as a collection of such systems. This possibility was first investigated by Davio and Deschamps [45], and subsequently by Deschamps [49] and Brown [27].

We say that a specification is *tabular* in case it is equivalent to a system of the form (9.28). A specification is tabular (as we shall show) if and only if it can be expressed by a truth-table. Nearly all current approaches to digital design are based therefore on tabular specifications.

To characterize tabular specifications, we note that (9.28) is equivalent to the system

$$
\begin{array}{ccc}
\phi_1(X, z_1) & = & 1 \\
\phi_2(X, z_2) & = & 1 \\
& \vdots & \\
\phi_n(X, z_n) & = & 1 ,
\end{array}
\tag{9.29}
$$

where

$$
\phi_i(X, z_i) = \alpha_i'(X)z_i' + \beta_i(X)z_i
\tag{9.30}
$$

for $i = 1, 2, \ldots, n$. Thus (9.28) is a general solution of the normal-form specification $\phi(X, Z) = 1$, where ϕ is related to the functions $\phi_1, \phi_2, \ldots, \phi_n$ of (9.29) as follows:

$$
\phi(X, Z) = \phi_1(X, z_1)\phi_2(X, z_2) \cdots \phi_n(X, z_n) .
\tag{9.31}
$$

We conclude that a specification equivalent to $\phi(X, Z) = 1$ is tabular if and only if there are switching functions $\phi_1, \phi_1, \ldots, \phi_n \colon \{0,1\}^{m+1} \longrightarrow \{0,1\}$ such that the multiplicative expansion (9.31) holds.

Theorem 9.3.1 *A specification equivalent to*

$$\phi(x_1, x_2, \ldots, x_m,, z_1, z_2, \ldots, z_n) = 1 \tag{9.32}$$

is tabular if and only if, for each $A \in \{0,1\}^m$, the discriminant $\phi(A, Z)$ is either zero or reduces to a term on the z-variables.

Proof.   Suppose that (9.32) is tabular, *i.e.*, suppose $\phi_1, \phi_1, \ldots, \phi_n$ exist such that the expansion (9.31) holds. It is clear that the condition $\phi_i(A, z_i) \in \{0, z_i', z_i, 1\}$ holds; hence, for each $A \in \{0,1\}^m$, $\phi(A, Z)$ is either zero or is a Z-term.

Suppose conversely that, for all $A \in \{0,1\}^m$, the discriminant $\phi(A, Z)$ is either zero or reduces to a Z-term. Define Boolean functions $\phi_i(A, z_i)$ for $A \in \{0,1\}^m$ and $i \in \{1, 2, \ldots, n\}$ as follows: if $\phi(A, Z)$ is zero, then $\phi_i(A, z_i) = 0$; if $\phi(A, Z)$ is a Z-term, then

> If z_i' is present in the term $\phi(A, Z)$, $\phi_i(A, z_i) = z_i'$.
> If z_i is present in the term $\phi(A, Z)$, $\phi_i(A, z_i) = z_i$.
> If neither z_i' nor z_i is present, $\phi_i(A, z_i) = 1$.

We may thus express $\phi(A, Z)$ as $\prod_{i=1}^{n} \phi_i(A, z_i)$, enabling us to develop $\phi(X, Z)$ in X-maxterms as follows:

$$
\begin{aligned}
\phi(X, Z) &= \prod_{A \in \{0,1\}^m} [\phi(A, Z) + (X^A)'] \\[2mm]
&= \prod_{A \in \{0,1\}^m} [\,[\prod_{i=1}^{n} \phi_i(A, z_i)] + (X^A)'] \\[2mm]
&= \prod_{A \in \{0,1\}^m} \prod_{i=1}^{n} [\phi_i(A, z_i) + (X^A)'] \\[2mm]
&= \prod_{i=1}^{n} \prod_{A \in \{0,1\}^m} [\phi_i(A, z_i) + (X^A)'] \\[2mm]
&= \prod_{i=1}^{n} \phi_i(X, z_i) \ .
\end{aligned}
$$

Thus (9.32) is tabular. $\square$

Example 9.3.1 Equation (9.12) is a tabular specification for J and K in terms of Q, R, S, and T, because each of the discriminants $\phi(J, K, 0, 0, 0, 0)$, $\phi(J, K, 0, 0, 0, 1)$, ..., $\phi(J, K, 1, 1, 1, 1)$ is either zero or reduces to a J,K-term. Equation (9.12) is not, however, a tabular specification for R, S, and T in terms of J, K, and Q; the discriminant $\phi(1, 0, 0, R, S, T)$, for example, evaluates to $R'S'T + R'ST'$, which does not reduce to a term on R,S,T. □

Example 9.3.2 Let the function ϕ in a normal-form specification be given by

$$\phi(X, Z) = \begin{array}{l} x_1'x_3'z_1z_2 + x_1x_2x_3 + x_1x_2z_1'z_2 + \\ + x_2'x_3'z_1z_2' + x_1x_2'z_1'z_2' + x_1x_3z_1' \end{array} \tag{9.33}$$

The discriminants of $\phi(X, Z)$ with respect to X are

$$\begin{array}{rcl}
\phi(0, 0, 0, Z) &=& z_1 \\
\phi(0, 0, 1, Z) &=& 0 \\
\phi(0, 1, 0, Z) &=& z_1z_2 \\
\phi(0, 1, 1, Z) &=& 0 \\
\phi(1, 0, 0, Z) &=& z_2' \\
\phi(1, 0, 1, Z) &=& z_1' \\
\phi(1, 1, 0, Z) &=& z_1'z_2 \\
\phi(1, 1, 1, Z) &=& 1 ,
\end{array} \tag{9.34}$$

each of which is either 0 or a term on Z. Thus the specification is tabular. □

Given a tabular specification $\phi(X, Z) = 1$, the functions $\phi_1, \phi_1, \ldots, \phi_n$ in the expansion (9.31) are not unique. The set (9.28) of intervals derived from these functions, however, and thus the set of particular solutions, is unique. A convenient set of functions is the one constructed in the proof of Theorem 9.3.1. These are found by the following rule: for $i \in \{1, 2, \ldots, n\}$,

$$\phi_i(X, z_i) = EDIS(\phi(X, Z), Z - \{z_i\}) , \tag{9.35}$$

If $Z = \{z_1, z_2, z_3\}$, for example, the functions

$$\begin{array}{rcl}
\phi_1(X, z_1) &=& EDIS(\phi(X, Z), \{z_2, z_3\}) \\
\phi_2(X, z_2) &=& EDIS(\phi(X, Z), \{z_1, z_3\}) \\
\phi_3(X, z_3) &=& EDIS(\phi(X, Z), \{z_1, z_2\})
\end{array} \tag{9.36}$$

x_1	x_2	x_3	z_1	z_2
0	0	0	1	X
0	1	0	1	1
1	0	0	X	0
1	0	1	0	X
1	1	0	0	1
1	1	1	X	X

Table 9.1: Sample Truth-table.

are functions suitable for the expansion (9.31).

Tabular specifications are precisely those that can be represented by a truth-table. To clarify the connection, let us examine Table 9.1.

We notice two kinds of don't-care specifications in Table 9.1. The first, represented by an absent row, is a forbidden input-combination; the second kind of don't-care, represented by an X in the table, denotes an output variable (corresponding to an input-combination that may occur) that may be freely assigned on $\{0, 1\}$.

Let us reduce the specification expressed by Table 9.1 to normal form. We begin by noting that the input-combinations forbidden by the table are $X = (0, 0, 1)$ and $X = (0, 1, 1)$. These prohibitions are represented by the system

$$
\begin{aligned}
x_1' x_2' x_3 &= 0 \\
x_1' x_2 x_3 &= 0 \, .
\end{aligned}
\tag{9.37}
$$

The six rows of the table are expressed by the implications

$$
\begin{aligned}
x_1' x_2' x_3' = 1 &\implies z_1 = 1 \\
x_1' x_2 x_3' = 1 &\implies z_1 z_2 = 1 \\
x_1 x_2' x_3' = 1 &\implies z_2' = 1 \\
x_1 x_2' x_3 = 1 &\implies z_1' = 1 \\
x_1 x_2 x_3' = 1 &\implies z_1' z_2 = 1 \\
x_1 x_2 x_3 = 1 &\implies 1 = 1
\end{aligned}
\tag{9.38}
$$

which are equivalent collectively to the system

$$
\begin{aligned}
x_1' x_2' x_3'(z_1)' &= 0 \\
x_1' x_2 x_3'(z_1 z_2)' &= 0 \\
x_1 x_2' x_3'(z_2')' &= 0 \\
x_1 x_2' x_3(z_1')' &= 0 \\
x_1 x_2 x_3'(z_1' z_2)' &= 0 \\
x_1 x_2 x_3(1)' &= 0
\end{aligned}
\tag{9.39}
$$

of equations. The system composed of (9.37) and (9.39) (and thus the original truth-table) is equivalent to the single equation

$$
\phi(X, Z) = 1 .
\tag{9.40}
$$

The function ϕ in (9.40) is given by

$$
\begin{aligned}
\phi(X, Z) &= x_1' x_2' x_3' z_1 + x_1' x_2 x_3' z_1 z_2 + x_1 x_2' x_3' z_2' + \\
&\quad x_1 x_2' x_3 z_1' + x_1 x_2 x_3' z_1' z_2 + x_1 x_2 x_3
\end{aligned}
\tag{9.41}
$$

in terms of the minterms of X; this is the same function as that shown in (9.33). There is a direct connection between truth-table (Table 9.1) and formula (above): each term of the formula specifies a row of the truth-table. This direct relationship is an advantage of standardizing on the 1-normal form for a specification.

9.4   Strongly Combinational Solutions

A system of the form

$$
Z = F(X, Z)
\tag{9.42}
$$

will be called an *implicit solution* of a specification $\mathcal{R}$ in case (9.42) implies $\mathcal{R}$. An implicit solution of $\mathcal{R}$ having the specialized form

$$
Z = \hat{F}(X)
\tag{9.43}
$$

will be called an *explicit solution* of $\mathcal{R}$.

Iterates.   Define the *iterates* $F^1, F^2, \ldots$ of $F(X, Z)$ in (9.42) as follows:

$$
\begin{aligned}
F^1(X, Z) &= F(X, Z) \\
F^{k+1}(X, Z) &= F(X, F^k(X, Z)) \quad k = 1, 2, \ldots
\end{aligned}
\tag{9.44}
$$

If there is an integer k such that the iterate $F^k(X, Z)$ depends only on X, then $F^j(X, Z) = F^k(X, Z)$ for $j > k$. It can be shown that the least such integer, if one exists, is less than 2^n. If no such integer exists, then the sequence $F^1, F^2, \ldots$ ultimately becomes cyclic, all of its members depending essentially on Z.

Example 9.4.1 The iterates of the system

$$
\begin{aligned}
z_1 &= x_1 z_2' \\
z_2 &= x_1 + z_1
\end{aligned}
\tag{9.45}
$$

are the following:

$$
F^0(X, Z) = \begin{bmatrix} x_1 z_2' \\ x_1 + z_1 \end{bmatrix}
$$

$$
F^1(X, Z) = \begin{bmatrix} x_1(x_1 + z_1)' \\ x_1 + (x_1 z_2') \end{bmatrix} = \begin{bmatrix} 0 \\ x_1 \end{bmatrix}
\tag{9.46}
$$

$$
F^j(X, Z) = F^1(X, Z) \qquad j = 2, 3, \ldots
$$

□

Let us suppose that a circuit is characterized by the implicit solution (9.42), and that $F^k(X, Z)$ depends only on X for some integer k. Then the circuit's outputs are guaranteed to stabilize, after the signal-waveform traverses k "loops" within the circuit, to values depending uniquely on the input-values. We thus define the implicit specification (9.42) to be *strongly combinational* in case the iterate $F^{2^n-1}(X, Z)$ is independent of Z.

Example 9.4.2 The sequence $F^1, F^2, \ldots$ shown in Figure 9.2 is based on the implicit specification F^1. This specification is strongly combinational because F^6—and therefore all subsequent iterates—is dependent only on X. □

9.5 Least-Cost Recursive Solutions

Given a consistent Boolean specification (9.9) for a multiple-output combinational circuit, we seek to design a realization that achieves economy by making use of outputs, as well as inputs, to produce outputs. Thus we seek

```
F1:    Z1 = Z2 Z3'+ X Z1 + X'Z3'
       Z2 = X'+ Z1'Z3 + Z2 Z3'
       Z3 = Z2'+ X Z1

F2:    Z1 = Z2 + X Z1
       Z2 = X'+ Z1'
       Z3 = X Z3'+ X Z1

F3:    Z1 = Z3 + Z2 + X'+ Z1
       Z2 = X'+ Z1'Z3 + Z1'Z2'
       Z3 = X Z2 + X Z1

F4:    Z1 = 1
       Z2 = X'+ Z1'Z2'
       Z3 = X Z3 + X Z1 + X Z2

F5:    Z1 = 1
       Z2 = X'+ Z1'Z2'Z3'
       Z3 = X

F6:    Z1 = 1
       Z2 = X'
       Z3 = X

F7:    Z1 = 1
       Z2 = X'
       Z3 = X
```

Figure 9.2: An Iterate-Sequence.

functions $f_1, f_2, \ldots, f_n$ such that the system

$$
\begin{aligned}
z_1 &= f_1(X, Z) \\
z_2 &= f_2(X, Z) \\
&\ \vdots \\
z_n &= f_n(X, Z)
\end{aligned}
\tag{9.47}
$$

satisfies the following requirements:

- it implies the specification (9.9);

- it is strongly combinational; and

- it minimizes some reasonable measure of cost.

The first requirement ensures that the implemented circuit is a solution of (*i.e.,* does the job specified by) (9.9). The second requirement is needed because a circuit corresponding to (9.47) may involve feedback-loops that lead to physical paradoxes or oscillation. Some properly-operating combinational circuits do include feedback-loops [90, 100, 133, 186]; we adopt a safe approach, however, by forbidding such loops.

A particular solution of the specification (9.9) is a system of the form (9.1) that implies (9.9). The set of all particular solutions of (9.9) may be represented by a *general solution* (*cf.* Chapter 6) expressed as a system

$$
\begin{aligned}
\alpha_1(X) &\leq u_1 \leq \beta_1(X) \\
\alpha_2(X, u_1) &\leq u_2 \leq \beta_2(X, u_1) \\
\alpha_3(X, u_1, u_2) &\leq u_3 \leq \beta_3(X, u_1, u_2) \\
&\ \vdots \\
\alpha_n(X, u_1, u_2, \ldots, u_{n-1}) &\leq u_n \leq \beta_n(X, u_1, u_2, \ldots, u_{n-1})
\end{aligned}
\tag{9.48}
$$

of recurrent subsumptions, where $(u_1, u_2, \ldots, u_n)$ is a permutation of the output-vector $(z_1, z_2, \ldots, z_n)$. Although the set of particular solutions represented by a general solution is unique, the form of a general solution (9.48) may vary widely from one permutation of the output-variables to another.

Every particular solution

$$
\begin{aligned}
u_1 &= \phi_1(X) \\
u_2 &= \phi_2(X) \\
&\ \vdots \\
u_n &= \phi_n(X)
\end{aligned}
\tag{9.49}
$$

of (9.9) (and nothing else) is produced from (9.48) as follows:

- choose ϕ_1 in the interval $\quad \alpha_1(X) \le \phi_1(X) \le \beta_1(X)$;

- choose ϕ_2 in the interval $\quad \alpha_2(X, \phi_1(X)) \le \phi_2(X) \le \beta_2(X, \phi_1(X))$;

- etc.

From a general solution (9.48), we may construct solutions of the form

$$
\begin{aligned}
u_1 &= f_1(X) \\
u_2 &= f_2(X, u_1) \\
&\vdots \\
u_n &= f_n(X, u_1, u_2, \ldots, u_{n-1}),
\end{aligned}
\tag{9.50}
$$

by independent selection of the functions $f_1, f_2, \ldots, f_n$ in the intervals displayed in (9.48).

We call an implicit solution of the form (9.50) *recursive*. Such a solution satisfies the first two of the requirements listed at the beginning of this section. It implies the specification (9.9) because it is a solution of that specification. It is physically realizable because its recursive structure corresponds to a circuit free of feedback-loops: output u_1 depends only on inputs; output u_2 depends only on inputs and u_1; output u_3 depends only on inputs, u_1 and u_2; etc.

The third of the requirements at the beginning of this section is that the system (9.50) should minimize some reasonable measure of cost, *i.e.*, a measure related to circuit-complexity. The measure chosen should also be relatively easy to derive from the form of a solution. Solutions will be expressed by means of SOP (sum-of-products) formulas; hence a reasonable cost-measure is *gate-input count*, *i.e.*, the number of inputs that would be supplied to gates if the solution were implemented in a two-level AND-to-OR circuit. (Input-signals and their complements are assigned zero cost.) Some formulas and their associated gate-input costs are shown below.

$$
\begin{array}{ll}
a'bc + ab' & \text{Cost} = 5 + 2 = 7 \\
a + b' & \text{Cost} = 0 + 2 = 2 \\
a' + bcd' & \text{Cost} = 3 + 2 = 5 \\
abc' & \text{Cost} = 3 + 0 = 3
\end{array}
$$

Each cost is indicated as the sum of two numbers. The first number is the total count of inputs to first-level AND-gates; the second number is the count of inputs to the second-level OR-gate. A one-literal term does not require an AND-gate; a one-term formula does not require an OR-gate. Although the gate-input cost is discussed in terms of gates, it is intended to measure the complexity of a collection of Boolean formulas and does not imply a specific implementation.

To clarify some of the foregoing ideas, let us consider an example.

Example 9.5.1 The system

$$
\begin{aligned}
z_1 &= x_1 + x_2' x_3' + x_2 x_3 \\
z_2 &= x_1' x_2 + x_1' x_3 \\
z_3 &= x_1' x_2 x_3
\end{aligned}
\tag{9.51}
$$

has a gate-input cost of $7 + 6 + 3 = 16$. This system is equivalent to (and is also the unique explicit solution of) a specification of the form $f(x_1, x_2, x_3, z_1, z_2, z_3) = 1$, where f is given by

$$
\begin{aligned}
f &= x_1' x_2' x_3' z_1 z_2' z_3' + x_1' x_2' x_3 z_1' z_2 z_3' + x_1' x_2 x_3' z_1' z_2 z_3' + \\
&\quad + x_1' x_2 x_3 z_1 z_2 z_3 + x_1 z_1 z_2' z_3' .
\end{aligned}
\tag{9.52}
$$

Choosing the "natural" permutation $(u_1, u_2, u_3) = (z_1, z_2, z_3)$ of the output-variables, a general solution of $f = 1$ is

$$
\begin{aligned}
x_1 + x_2' x_3' + x_2 x_3 &\leq z_1 \leq x_1 + x_2' x_3' + x_2 x_3 \\
x_1' x_2' x_3 z_1' + x_1' x_2 x_3' z_1' + x_1' x_2 x_3 z_1 &\leq z_2 \leq x_1' x_2 + x_1' x_3 + z_1' \\
x_1' x_2 x_3 z_1 z_2 &\leq z_3 \leq z_1' z_2' + z_1 z_2 + x_2' x_3' z_1' + x_1 z_1' \\
&\qquad x_1' x_2 x_3 + x_1' x_3 z_1 + x_1' x_2 z_1 .
\end{aligned}
\tag{9.53}
$$

A large number of recursive particular solutions, all reducible to the unique explicit solution (9.51), can be derived from (9.53). Among the simplest of these is

$$
\begin{aligned}
z_1 &= x_1 + x_2' x_3' + x_2 x_3 \\
z_2 &= x_1' x_2 + z_1' \\
z_3 &= z_1 z_2 ,
\end{aligned}
\tag{9.54}
$$

for which the cost $7 + 4 + 2 = 13$.

The permutation $(u_1, u_2, u_3) = (z_2, z_3, z_1)$ leads to a general solution for which a simplified recursive solution is

$$
\begin{aligned}
z_2 &= x_1' x_2 + x_1' x_3 \\
z_3 &= x_1' x_2 x_3 \\
z_1 &= z_2' + z_3 ,
\end{aligned}
\tag{9.55}
$$

with an associated cost of $6 + 3 + 2 = 11$, a savings of 5 gate-inputs over the original explicit solution (9.51). $\square$

9.6   Constructing Recursive Solutions

Our objective is to find a least-cost recursive solution of a specification

$$\phi(X, Z) = 1 . \tag{9.56}$$

Let z_i be an argument in Z and denote the arguments in Z other than z_i by Z_{-i}, i.e.,

$$Z_{-i} = Z - \{z_i\} ,$$

and let V be a subset of Z_{-i}. Given the specification (9.56), we associate three sets, $R(z_i)$, $S(z_i, V)$, and $T(z_i)$, as follows with z_i and V:

$$
\begin{aligned}
R(z_i) &= [(\phi/z_i')', (\phi/z_i)] & (9.57) \\
S(z_i, V) &= [(EDIS(\phi, V)/z_i')', EDIS(\phi, V)/z_i] & (9.58) \\
T(z_i) &= [(\phi/z_i')' \cdot (\phi/z_i), (\phi/z_i')' + (\phi/z_i)] . & (9.59)
\end{aligned}
$$

A subsumptive general solution of the specification, if produced by the method of successive eliminations, defines the set of allowable z_i by

$$z_i \in S(z_i, V) , \tag{9.60}$$

the subset V being determined by the permutation of $(z_1, z_2, \ldots, z_n)$ employed in constructing the solution. Example 9.5.1 illustrates that the forms (and therefore the costs) of recursive solutions are dependent upon that permutation. One approach to finding a least-cost recursive solution, therefore, is to construct a general solution corresponding to each of the $n!$ permutations of the output-variables, to determine a least-cost solution based on each, and to select the best of such solutions. We describe an alternative approach, based on $T(z_i)$ rather than on $S(z_i, V)$.

Lemma 9.6.1 *If (9.56) is a tabular specification, $z_i \in Z$, and $V \subseteq Z_{-i}$, then*

$$R(z_i) \subseteq S(z_i, V) \subseteq T(z_i) . \tag{9.61}$$

Proof.   It follows from the tabular property of (9.56) that there are switching functions g and h such that

$$\phi(X, Z) = g(X, z_i) \cdot h(X, Z_{-i}) , \tag{9.62}$$

whence

$$(\phi/z_i) \;=\; g(X,1)\cdot h(X,Z_{-i}) \tag{9.63}$$
$$EDIS(\phi,V)/z_i \;=\; g(X,1)\cdot EDIS(h(X,Z_{-i}),V)\,. \tag{9.64}$$

Theorem 4.8.4 guarantees that $h(X,Z_{-i}) \le EDIS(h(X,Z_{-i}),V)$; thus

$$(\phi/z_i) \le EDIS(\phi,V)/z_i\,.$$

The relationship

$$(\phi/z_i')' + (\phi/z_i) = g'(X,0) + g(X,1) + h'(X,Z_{-i}) \tag{9.65}$$

follows from (9.62), and thus the inclusion

$$EDIS(\phi,V)/z_i \le (\phi/z_i')' + (\phi/z_i) \tag{9.66}$$

is verified on comparison of (9.64) and (9.65). We have thus verified the first inclusion-pair of the system

$$(\phi/z_i) \;\le\; EDIS(\phi,V)/z_i \;\le\; (\phi/z_i')' + (\phi/z_i) \tag{9.67}$$
$$(\phi/z_i')' \cdot (\phi/z_i) \;\le (EDIS(\phi,V)/z_i')' \le\; (\phi/z_i')'\,. \tag{9.68}$$

The second inclusion-pair, (9.68), is verified by similar reasoning. $\square$

Example 9.6.1 Given the tabular specification

$$x_1'x_2'z_1' + x_1'x_2z_2' + x_1x_2'z_1z_2z_3' = 1\,, \tag{9.69}$$

the intervals $\mathcal{R}(z_1)$ and $\mathcal{T}(z_1)$ are given by

$$\mathcal{R}(z_1) \;=\; [x_1 + x_2z_2 \quad,\quad x_1x_2'z_2z_3' + x_1'x_2z_2']$$
$$\mathcal{T}(z_1) \;=\; [x_1x_2'z_2z_3' \quad,\quad x_1 + x_2]\,.$$

The interval $\mathcal{S}(z_1,V)$, for various subsets V, is listed below:

$$\mathcal{S}(z_1,\{z_2\}) \;=\; [x_1 \qquad\quad,\quad x_1x_2'z_3' + x_1'x_2]$$
$$\mathcal{S}(z_1,\{z_3\}) \;=\; [x_1 + x_2z_2 \quad,\quad x_1x_2'z_2 + x_1'x_2z_2']$$
$$\mathcal{S}(z_1,\{z_2,z_3\}) \;=\; [x_1 \qquad\quad,\quad x_1x_2' + x_1'x_2]\,.$$

Comparisons among the foregoing intervals verify the inclusions in (9.61). $\square$

Theorem 9.6.1 *Let (9.56) be a tabular specification and let z_i be an argument in Z. If the arguments in $X \cup Z_{-i}$ are constrained so that (9.56) can be solved for z_i, then*

$$\mathcal{R}(z_i) = \mathcal{S}(z_i, V) = \mathcal{T}(z_i) . \tag{9.70}$$

Proof. Equation (9.56) can be solved for z_i if and only if the consistency-condition

$$EDIS(\phi, \{z_i\}) = 1 \tag{9.71}$$

holds. Assume that the arguments in $X \cup Z_{-i}$ are constrained so that (9.71), *i.e.*,

$$(\phi/z_i') + (\phi/z_i) = 1 , \tag{9.72}$$

is satisfied, whence the equations

$$(\phi/z_i')' = (\phi/z_i')' \cdot (\phi/z_i) \tag{9.73}$$
$$(\phi/z_i) = (\phi/z_i')' + (\phi/z_i) \tag{9.74}$$

become identities. The set-equality

$$\mathcal{R}(z_i) = \mathcal{T}(z_i) \tag{9.75}$$

therefore holds, whence (9.70) follows from Lemma 9.6.1. $\square$

Neither Lemma 9.6.1 nor Theorem 9.6.1 holds for a non-tabular specification, as the following example shows.

Example 9.6.2 The specification $\phi(x_1, x_2, z_1, z_2) = 1$, where ϕ is given by

$$\phi = x_1' z_1 + x_1' x_2' z_2 + x_1 z_1' z_2 , \tag{9.76}$$

is non-tabular: $\phi(0, 0, z_1, z_2)$, for example, evaluates to $z_1 + z_2$, which is not reducible to a term on (z_1, z_2). The interval $\mathcal{S}(z_1, \{z_2\})$ is given by

$$\begin{aligned} \mathcal{S}(z_1, \{z_2\}) &= [(EDIS(\phi, \{z_2\})/z_i')', \ EDIS(\phi, \{z_2\})/z_i] \\ &= [x_1' x_2, \ x_1'] . \end{aligned}$$

The interval $\mathcal{T}(z_i)$, on the other hand, is given by

$$\mathcal{T}(z_i) = [x_1' x_2 + x_1' z_2', \ x_1' + z_2'] . \tag{9.77}$$

The element $x_1' x_2$ belongs to $\mathcal{S}(z_1, \{z_2\})$ but not to $\mathcal{T}(z_i)$; hence $\mathcal{S}(z_1, \{z_2\})$ is not a subset of $\mathcal{T}(z_i)$. $\square$

9.6.1   The Procedure

The following procedure produces a least-cost recursive solution of a consistent tabular specification $\phi(X, Z) = 1$.

1. For each $z_i \in Z$, calculate $T(z_i)$ by use of the relation (9.59), and determine the minimal determining subsets (*cf.* Section 4.9) on $T(z_i)$. A minimal determining subset, $\mathcal{M}$, on $T(z_i)$ is a subset of $X \cup Y$ having the following properties:

   (a) the arguments of $\mathcal{M}$ suffice to express at least one function in $T(z_i)$; and

   (b) none of the proper subsets of $\mathcal{M}$ has property (a).

2. Assign a cost, $c(\mathcal{M})$, to each minimal determining subset $\mathcal{M}$ found in Step 2. Examples of possible cost-measures are

   - $c(\mathcal{M}) = $ the number of arguments comprised by $\mathcal{M}$.
   - $c(\mathcal{M}) = $ the cost of a formula, expressed by the arguments in $\mathcal{M}$ and representing a function in the set $T(z_i)$, that has least cost over all functions in $T(z_i)$.

3. Select a minimal determining subset, call it $\mathcal{M}(z_i)$, corresponding to each $z_i \in Z$. This selection should meet the following conditions:

   (a) There is a permutation $(u_1, u_2, \ldots, u_n)$ of $(z_1, z_2, \ldots, z_n)$ such that
   - $\mathcal{M}(u_1) \subseteq X$, and
   - $\mathcal{M}(u_k) \subseteq X \cup \{u_1, \ldots, u_{k-1}\}$ $(1 \leq k \leq n)$.

   (b) The total cost, *i.e.*,
   $$\sum_{z_i \in Z} c(\mathcal{M}(z_i)) ,$$
   is minimized over all permutations (a).

This procedure avoids the construction of a general solution, and subsequent location of a least-cost recursive solution, for each permutation of Z. Instead, the minimal determining subsets (and their associated costs) are determined at the outset for each variable z_i. Only the contents and costs of the minimal determining subsets are then required to complete the procedure.

The procedure has a number of limiting characteristics, summarized as follows:

Example 9.6.3 The following example-specification was given in an early text [82, p. 90] on digital design:

$$
\begin{aligned}
z_1 &= x_1'x_2x_3 + x_1x_2'x_3' + x_1x_2'x_3 + x_1x_2x_3' \\
z_2 &= x_1'x_2x_3 + x_1x_2'x_3 + x_1x_2x_3' \\
z_3 &= x_1'x_2'x_3' + x_1'x_2x_3 + x_1x_2'x_3 + x_1x_2x_3' \,.
\end{aligned}
\tag{9.78}
$$

The transformed system shown in [82] is

$$
\begin{aligned}
z_1 &= x_1'x_2x_3 + x_1x_2' + x_1x_3' \\
z_2 &= x_2z_1 + x_3z_1 \\
z_3 &= x_1'x_2'x_3' + z_2 \,,
\end{aligned}
\tag{9.79}
$$

which has a gate-input cost of 21. The argument-set used to compute z_3, viz., $\{x_1, x_2, x_3, z_2\}$, is not minimal; one of its proper subsets, $\{x_1, x_2, x_3\}$, clearly suffices to determine z_3. A least-cost system based solely on minimal determining subsets is

$$
\begin{aligned}
z_1 &= x_1'x_2x_3 + x_1x_2' + x_1x_3' \\
z_2 &= z_1z_3 \\
z_3 &= x_3z_1 + x_2z_1 + x_2'x_3'z_1' \,,
\end{aligned}
\tag{9.80}
$$

which has a gate-input cost of 22. $\square$

9.6.2 An Implementation using BORIS

A program has been written using the reasoning-toolset BORIS (Boolean Reasoning In Scheme) to construct a least-cost recursive solution for a tabular specification.

The program accepts a system of Boolean equations or a representation of an incompletely-specified truth-table as input; either format is reduced by the program to a specification of the form $\phi = 1$, where ϕ is a Boolean function. After printing the terms of an SOP formula for ϕ, the program determines and prints the minimal z_i-determining subsets for each output-variable z_i. These subsets are the basis for the subsequent search for an ordering of the output-variables leading to a least-cost recursive solution. Partial permutations are built up during the search-process; the first output-variable in such a partial permutation, u_1, must depend only on input-variables. The next output-variable, u_2, may depend on u_1 as well as on inputs; u_3 may depend on u_1 and and u_2 as well as on inputs, and so on. The search-process is branch-and-bound, maintaining open and closed sets of partial permutations, using gate-input count (discussed in Section 9.5) as a measure of cost. In the example discussed below, the sequence of best partial permutations generated during the search is

$$(8 \ (V \ 8 \ A \ C \ D))$$
$$(11 \ (W \ 11 \ A \ B \ C \ D))$$
$$(14 \ (V \ 8 \ A \ C \ D) \ (U \ 6 \ B \ V))$$
$$(15 \ (V \ 8 \ A \ C \ D) \ (W \ 7 \ A \ B \ V))$$
$$(15 \ (U \ 15 \ A \ B \ C \ D))$$

A partial permutation is represented, as shown above, by a list of the form

$$(\text{Cost} \ P(u_1) \ \ldots \ P(u_k)) \, .$$

where "Cost" denotes total accumulated gate-input cost and each sublist $P(u_j)$ $(j = 1, \ldots, k)$ has the form

$$(u_j \ c(\mathcal{M}(u_j)) \ \mathcal{M}_1(u_j) \ \mathcal{M}_2(u_j) \ \ldots) \, .$$

The list

$$(14 \ (V \ 8 \ A \ C \ D) \ (U \ 6 \ B \ V))$$

represents a typical partial permutation. This shows that V is the first output-variable in the permuted sequence, with functional dependence on variables A, C, D, and gate-input cost 8. The second output-variable in the sequence is U, with dependence on B and V, and cost 6. The total cost, 8 + 6 = 14, is displayed first in the list.

Example 9.6.4 A circuit is to be designed in conformity with the 14-row truth-table shown in Table 9.2. The inputs are a, b, c, and d; the outputs are u, v, and w. Using standard design-techniques, taking advantage of the "don't-care" entries in the table, a least-cost realization is

$$
\begin{aligned}
u &= bc + bd + a'cd + a'b'c'd' \\
v &= a'cd + a'c'd' \\
w &= a + b'c + b'd + bc'd'
\end{aligned}
$$

with costs for u, v, and w of 15, 8, and 11, respectively. Thus the least cost using conventional techniques is 34.

a	b	c	d	u	v	w
0	0	0	0	1	1	0
0	0	0	1	0	0	1
0	0	1	0	0	0	1
0	0	1	1	1	1	X
0	1	0	0	0	1	1
0	1	0	1	1	0	0
0	1	1	0	1	0	0
0	1	1	1	X	X	X
1	0	0	0	0	0	1
1	0	0	1	X	X	1
1	0	1	0	X	0	1
1	0	1	1	0	0	X
1	1	0	0	X	0	X
1	1	0	1	1	0	1

Table 9.2: Truth-table for sample design.

Figure 9.3 shows the BORIS-output in designing a recursive realization of the circuit. The Scheme-function DESIGN has two arguments: the first, named SAMPLE in this case, denotes a specification, $\mathcal{R}$ (a truth-table in this example); the second argument denotes the output-vector, Z. The result,

$$v = a'cd + a'c'd'$$
$$u = b'v + bv'$$
$$w = u + a',$$

has a gate-input cost of 16.

[10] (DESIGN SAMPLE '(U V W))

Function:
A'B'C'D'U V W'
A'B'C'D U'V'W
A'B'C D'U'V'W
A'B'C D U V
A'B C'D'U'V W
A'B C'D U V'W'
A'B C D'U V'W'
A'B C D
A B'C'D'U'V'W
A B'C'D W
A B'C D'V'W
A B'C D U'V'
A B C'D'V'
A B C'D U V'W

Minimal Determining Subsets:
U ((B V) (A B C D) (A C D W))
V ((A B U) (A C D))
W ((A U) (A B V) (A B C D))

(0)
(8 (V 8 A C D))
(11 (W 11 A B C D))
(14 (V 8 A C D) (U 6 B V))
(15 (V 8 A C D) (W 7 A B V))
(15 (U 15 A B C D))

(16 (V 8 A C D) (U 6 B V) (W 2 A U))
   U = B'V + B V'
   V = A'C D + A'C'D'
   W = U'+ A

DONE

Figure 9.3: BORIS-output for sample design.

Appendix A

Syllogistic Formulas

Our approach to Boolean reasoning owes much to the work of A. Blake [10]. In this Appendix we outline Blake's theory of syllogistic Boolean formulas, modifying his notation and some details of his proofs, but retaining insofar as possible his point of view.

The reader is assumed to be familiar with the definitions given at the beginning of Chapter 4 concerning Boolean formulas; some definitions, however, are repeated for convenience. We assume that Boolean functions are expressed by disjunctive normal (SOP) formulas; thus "formula" will invariably mean "disjunctive normal formula." A Boolean function will be denoted by one of the lower-case letters f, g, h and a formula representing that function by the corresponding upper-case letter (F, G, or H). A term (conjunct) will be represented by one of the lower-case letters p, q, r, s, t; a term will be treated either as a function or as a formula, depending on context. Literals are denoted by x, y, or z.

Two formulas will be called *equivalent* ($\equiv$) in case they represent the same function, i.e., in case one can be transformed into the other, in a finite number of steps, by application of the laws of Boolean algebra. Two formulas will be called *congruent* ($\overset{\circ}{=}$) in case one can be transformed into the other using only the commutative law. Thus congruent formulas may differ only in the order of enumeration of their terms and in the order of the literals comprised by any term.

Given two Boolean functions g and h, we say that g is *included in* h, written $g \leq h$, in case the identity $gh' = 0$ is fulfilled. When applied to formulas (e.g., $G \leq H$), the relation $\leq$ will refer to the functions those formulas represent.

A.1   Absorptive Formulas

An SOP formula F will be called *absorptive* in case no term in F is absorbed by any other term in F. If F is not absorptive, then an equivalent absorptive formula, which we call $ABS(F)$, may be obtained from F by successive deletion of terms absorbed by other terms in F.

Lemma A.1.1 *The formula $ABS(F)$ is unique to within congruence.*

Proof. Suppose G_1 and G_2 are two absorptive formulas derived from F by the deletion, in different order, of absorbed terms. Let p be a term of G_1. Then p is a term of F that is not absorbed by any other distinct term of F; hence, p must be a term of G_2. Similarly, any term of G_2 must be a term of G_1. Hence, $G_1 \overset{\circ}{=} G_2$. $\square$

It is clear that $ABS(F)$ is equivalent to F. There may be absorptive formulas equivalent to F, however, that are not congruent to $ABS(F)$. Let F, for example, be the formula

$$ac' + b'c + a'b + a'b'c .$$

Then $ABS(F)$ is the formula $ac' + b'c + a'b$. The absorptive formula

$$a'c + bc' + ab'$$

is equivalent to F, but not congruent to $ABS(F)$.

A.2   Syllogistic Formulas

Let F and G be SOP formulas. We say that G is *formally included* in F, written $G \ll F$, in case each term of G is included in some term of F. We write $G \not\ll F$ if G is not formally included in F. Formal inclusion clearly implies inclusion, i.e., $G \ll F \Longrightarrow G \leq F$ for any F, G pair. Formula F will be called *syllogistic* in case the converse also holds, i.e., in case, for every SOP formula G,

$$G \leq F \Longrightarrow G \ll F .$$

Thus F is syllogistic if and only if every implicant of F is included in some term of F.

Lemma A.2.1 *Let F, G, and H be SOP formulas. If $F \ll G + H$ and $G \ll H$, then $F \ll H$.*

Proof. Consider any term p of F, and suppose that $p \not\ll H$. Then there is a term q of G such that $p \leq q$. Since $G \ll H$, there is a term r of H such that $q \leq r$. Thus $p \leq r$, whence $p \ll H$, a contradiction. Thus every term of F is formally included in H. $\square$

Lemma A.2.2 *Let F be an SOP formula. F is syllogistic if and only if $ABS(F)$ is syllogistic.*

Proof. Suppose F is syllogistic and let p be an implicant of $ABS(F)$. Then $p \leq F$, whence $p \ll F$, i.e., there is a term q of F such that $p \leq q$. Let r be a maximal term of F (i.e., a term made up of a minimal number of letters), possibly q, such that $q \leq r$. Now $p \leq r$ and r must be a term of $ABS(F)$; therefore $p \leq ABS(F)$ and we conclude that $ABS(F)$ is syllogistic. Suppose, conversely, that $ABS(F)$ is syllogistic. Every term of $ABS(F)$ is a term of F; hence F must also be syllogistic. $\square$

Lemma A.2.3 *Let F_1 and F_2 be syllogistic. If $F_1 \equiv F_2$ then $ABS(F_1) \stackrel{\circ}{=} ABS(F_2)$.*

Proof. Suppose F_1 and F_2 to be equivalent syllogistic formulas. We deduce from Lemma A.2.2 that $ABS(F) \ll ABS(G)$ and that $ABS(G) \ll ABS(F)$. Let p be a term of $ABS(F)$. There is a term q of $ABS(G)$ such that $p \leq q$; also, there is a term r of $ABS(F)$ such that $q \leq r$. Thus $p \leq r$, whence $p = r$ (because $ABS(F)$ is absorptive) and therefore $p = q$. We conclude that every term of $ABS(F)$ is a term of $ABS(G)$; similarly, every term of $ABS(G)$ is a term of $ABS(F)$. Hence, $ABS(F) \stackrel{\circ}{=} ABS(G)$. $\square$

Given SOP formulas F and G, we define $F \times G$ to be the SOP formula produced by multiplying out the conjunction FG, using the distributive laws. If $F = \sum_i s_i$ and $G = \sum_j t_j$, then

$$F \times G = \sum_i \sum_j s_i \cdot t_j \,,$$

where repeated literals are dropped in each product $s_i \cdot t_j$ of terms, $s_i \cdot 1 = s_i$, and $1 \cdot t_j = 1$; also a product is dropped if it contains a complementary pair of literals. The operation $\times$ is commutative and associative; hence, $F_1 \times F_2 \times \cdots \times F_k$ denotes without ambiguity the SOP formula produced by multiplying out $F_1 F_2 \cdots F_k$ in the manner discussed above.

Theorem A.2.1 *Let $F_1, \ldots, F_k$ be syllogistic formulas. Then $F_1 \times \cdots \times F_k$ is syllogistic.*

Proof. Let t be an implicant of $F_1 \times \cdots \times F_k$. Then $t \leq F_i$ for $i = 1, 2, \ldots, k$; further, $t \ll F_i$, since the F_i are syllogistic. Thus each of the F_i contains a term p_i such that $t \leq p_i$, and therefore $t \leq \prod_{i=1}^{k} p_i$. But $\prod_{i=1}^{k} p_i$ is a term of $F_1 \times \cdots \times F_k$; hence $F_1 \times \cdots \times F_k$ is syllogistic. $\square$

Let a be any letter. Two terms will be said to have an *opposition* in case one term contains the literal a and the other the literal a'. (If the symbol x stands for the literal a', then we shall understand x' to stand for a.)

Lemma A.2.4 *If terms r and s have no oppositions, then $r+s$ is syllogistic.*

Proof. We assume that neither r nor s is the term 1, for which case the lemma holds trivially. Suppose the lemma to be false, i.e., suppose that there are terms r and s having no oppositions such that $r + s$ is not syllogistic. Then there is a term t such that $t \leq r+s$, $t \nleq r$, and $t \nleq s$. Thus each of the terms r and s contains a literal not in t, i.e., $r = xp$ and $s = yq$, where x and y are literals not in t, p is a term not involving x, and q is a term not involving y. Now $t \leq r + s \Longrightarrow tr's' = 0 \Longrightarrow t(x' + p')(y' + q') = 0 \Longrightarrow tx'y' = 0$. Thus, either $x'y' = 0$ or one of the literals x or y appears in t. The former is ruled out by the hypothesis that r and s have no oppositions, the latter by explicit assumption; hence, we have arrived at a contradiction. $\square$

Theorem A.2.2 *Let r and s be terms. The formula $r + s$ is non-syllogistic if and only if r and s have exactly one opposition.*

Proof. Let k be the number of oppositions between r and s. If $k = 0$, then $r + s$ is syllogistic by Lemma A.2.4. Suppose $k \geq 1$, i.e., suppose $r = x'p$ and $s = xq$, where x is a literal and p and q are terms not involving x' or x (if $r = x'$, then $p = 1$; if $s = x$, then $q = 1$). Consider first $k = 1$, in which case $pq \neq 0$. Let t be the term formed from pq by deleting duplicate literals. Then $t \leq r + s$, since $tr's' = pq(x + p')(x' + q') = 0$. However $t \nleq r$, because $tr' = pq(x + p') = pqx \neq 0$. It follows similarly that $t \nleq s$. Thus $r + s$ is not syllogistic if $k = 1$. Consider now $k > 1$, in which case $pq = 0$, and let t be any term such that $t \leq r+s$, so that $tr's' = t(x+p')(x'+q') = txq'+tx'p' = 0$. Then $tx \leq q$ and $tx' \leq p$, from which we deduce that $tx \leq qx = s$ and $tx' \leq px' = r$. Either x or x' must appear in t, for suppose neither appears. Then $txq' + tx'p' = 0 \Longrightarrow tq' + tp' = 0 \Longrightarrow t \leq pq$. But $pq = 0$ for $k > 1$; hence $t = 0$, contradicting the assumption that t is a term. If x appears in t, then $tx = t$ and therefore $t \leq s$. If x' appears in t, then $tx' = t$ and therefore $t \leq r$. If $k > 1$, therefore, $t \leq r + s$ implies that either $t \leq r$ or

$t \leq s$ for every term t, i.e., $r + s$ is syllogistic. We conclude that $r + s$ is non-syllogistic if $k = 1$ and is syllogistic otherwise. $\square$

Suppose two terms r and s have exactly one opposition. Then the *consensus* [161] of r and s, which we shall denote by $c(r, s)$, is the term obtained from the conjunction rs by deleting the two opposed literals as well as any repeated literals. The consensus $c(r, s)$ does not exist if the number of oppositions between r and s is other than one. The consensus of two terms was called their "syllogistic result" by Blake.

Lemma A.2.5 *Let $r + s$ be a non-syllogistic SOP formula. Then*

$$\begin{array}{ll} \text{(i)} & r + s + c(r, s) \equiv r + s \\ \text{(ii)} & r + s + c(r, s) \text{ is syllogistic.} \end{array}$$

Proof. Applying Theorem A.2.2, $r + s$ is non-syllogistic if and only if $r = x'p$ and $s = xq$, where p and q are terms such that $pq \neq 0$. The consensus $c(r, s)$ is the term formed from pq by deleting duplicate literals; let pq henceforth denote that term. To prove (i), we re-express $r + s + c(r, s)$ as $x'p + xq + pq$, which is equivalent, by Property 8, Section 3.5, to $x'p + xq$. To prove (ii), we show that if a term t is such that $t \leq r + s$ and $t \not\ll r + s$, then $t \leq pq$ (recalling that $c(r, s) = pq$). The condition $t \leq r + s$ holds if and only if $tr's' = txq' + tx'p' = 0$. Now t cannot involve x, for otherwise $txq' = 0 \implies tx(q' + x') = 0 \implies txs' = 0 \implies ts' = 0 \implies t \leq s$. Similarly, t cannot involve x'. Thus $txq' + tx'p' = 0 \implies tq' + tp' = t(pq)' = 0 \implies t \leq pq$. $\square$

Theorem A.2.3 *If an SOP formula F is not syllogistic, it contains terms p and q, having exactly one opposition, such that $c(p, q)$ is not formally included in F.*

Proof. Let n be the number of distinct letters appearing in F and define R to be the set of implicants of F that are not formally included in F. Define the *degree* of any member of R to be the number of its literals. Let t be any member of R of maximal degree; this degree is less than n because a term of degree n (i.e., a minterm) is formally included in any SOP formula in which it is included. There is therefore some letter, x, that appears in F but is absent from t. The terms tx' and tx are implicants of F whose degree is higher than that of t; hence, $tx' \ll F$ and $tx \ll F$, i.e., F contains terms p and q such that $tx' \leq p$ and $tx \leq q$; hence $t \leq p + q$. But t is not formally included in

$p + q$ and thus $p + q$ is not syllogistic; from Theorem A.2.2, therefore, p and q have exactly one opposition. From part (ii) of Lemma A.2.5, moreover, $t \leq c(p,q)$. Suppose $c(p,q) \ll F$; then $t \ll F$. But $t \not\ll F$ because t is a member of R. Hence $c(p,q) \not\ll F$. $\square$

Corollary A.2.1 *If an SOP formula F is not syllogistic, then $ABS(F)$ contains terms p and q, having exactly one opposition, such that $c(p,q) \not\ll ABS(F)$.*

Proof.   By Lemma A.2.3, if F is not syllogistic, then $ABS(F)$ is not syllogistic; hence Theorem A.2.3 is applicable to $ABS(F)$. $\square$

A.3   Prime Implicants

An *implicant* of a Boolean function f is a term p such that $p \leq f$. A *prime implicant* of f is an implicant of f that ceases to be so if any of its literals is removed. The concept of a prime implicant (due to Quine [161]) does not appear in Blake's development; however, prime implicants are intimately related, as we show, to syllogistic formulas.

Lemma A.3.1 *An implicant p of a Boolean function f is a prime implicant of f in case the implication*

$$p \leq q \leq f \quad \Longrightarrow \quad p = q \tag{A.1}$$

holds for every term q.

   Proof.   Suppose that p is an implicant of f satisfying (A.1) and that p is not a prime implicant of f. Then p is congruent to one of the forms xr or $x'r$, where x is a literal and r is an implicant of f, i.e., $r \leq f$. Thus $p \leq r \leq f$ and $p \neq r$, and we conclude that p does not satisfy (A.1), which is a contradiction; thus p is a prime implicant of f. Suppose on the other hand that p is a prime implicant of f, i.e., that $p \leq f$ and that if r is a proper subproduct of p, then $r \not\leq f$. Suppose further that $p \leq q \leq f$ for some term q. The condition $p \leq q$ holds between terms if and only if either $p = q$ or q is a proper subproduct of p. The latter is ruled out because no proper subproduct of a prime implicant of f is an implicant of f, and we have assumed that $q \leq f$. Hence $p = q$, establishing condition (A.1). $\square$

Lemma A.3.2 *If r is an implicant of f, then there is a prime implicant p of f such that $r \leq p$.*

Proof. If r is a prime implicant of f, then $p = r$. If r is not a prime implicant of f, then there is an implicant $q_1 \neq r$ of f such that $t \leq q_1 \leq f$. If q_1 is not a prime implicant of f, then there is an implicant $q_2 \neq q_1$ of f such that $q_1 \leq q_2 \leq f$. This process must ultimately terminate, yielding a prime implicant p of f such that $t \leq p$. $\square$

Theorem A.3.1 *Let F be an SOP formula for a Boolean function f. Then F is syllogistic if and only if every prime implicant of f is a term of F.*

Proof. Suppose F is syllogistic and let p be a prime implicant of f. Then $p \leq f$, whence $p \ll F$, i.e., $p \leq q \leq F$, where q is a term of F. Thus $p = q$ by the definition of a prime implicant, whence p is a term of F. Suppose on the other hand that every prime implicant of f is a term of F. Let t be a term such that $t \leq F$; by Lemma A.3.2 there is a prime implicant p of f (possibly t) such that $t \leq p$. But p is a term of F, and therefore $t \ll F$. Thus F is syllogistic. $\square$

A.4 The Blake Canonical Form

Let F be a syllogistic formula for a Boolean function f. We call the formula ABS(F) the *Blake canonical form* for f, and we denote it by $BCF(f)$. The function f determines the formula $BCF(f)$, by Lemma A.2.2, to within congruence. Blake called this formula the "simplified canonical form" and showed that it is minimal within any class of syllogistic formulas for f, i.e., if F is syllogistic, then $F \equiv BCF(f)$ implies that every term of $BCF(f)$ is a term of F.

Theorem A.4.1 *Let f be a Boolean function. Then $BCF(f)$ is the disjunction of all of the prime implicants of f.*

Proof. $BCF(f)$ is syllogistic (Lemma A.2.1); hence, by Theorem A.3.1, every prime implicant of f is a term of $BCF(f)$. It only remains to show that every term of $BCF(f)$ is a prime implicant of f. Suppose the contrary, i.e., suppose there is a term p of $BCF(f)$ that is not a prime implicant of f. From the relation $p \leq BCF(f)$ it follows that there is a term $q \neq p$ such that $p \leq q \leq BCF(f)$. Since $BCF(f)$ is syllogistic, $q \ll BCF(f)$, i.e., $BCF(f)$ contains a term r such that $q \leq r$. Thus $BCF(f)$ has distinct terms p and r such that $p \leq r$, which is a contradiction because $BCF(f)$ is absorptive. $\square$

Bibliography

[1] Adam, A., "An application of truth-functions in formalized diagnostics," *Acta Cybernetica*, vol. 2, pp. 291-298, 1976.

[2] Akers, S.B., "On a theory of Boolean functions," *J. Soc. Indust. Appl. Math.*, vol. 7, no. 4, pp. 487-498, Dec. 1959.

[3] Arnold, B.H., *Logic and Boolean Algebra*. Englewood Cliffs, N.J.: Prentice-Hall, 1962.

[4] Ashenhurst, R.L., "Simultaneous equations in switching theory," Report BL-5, Harvard Computation Lab., Harvard University, 1954, pp. 1-8.

[5] Ashenhurst, R.L., "The decomposition of switching functions," *Proc. International Symposium on the Theory of Switching,* April, 1957. Vol. 29 of *Annals of the Computation Laboratory of Harvard University,* pp. 74-116, 1959 (Included in [42] as an appendix).

[6] Beatson, T.J., "Minimization of components in electronic switching circuits," *Trans. A.I.E.E., Part I, Communications and Electronics*, vol. 77, pp. 283-291, 1958.

[7] Bennett, A.A. and C.A. Baylis, *Formal Logic: A Modern Introduction.* New York: Prentice-Hall, 1939.

[8] Bing, K., "On simplifying propositional formulas" (abstract) *Bull. Amer. Math. Soc.*, vol. 61, p. 560, 1955.

[9] Bing, K., "On simplifying truth-functional formulas," *J. Symbolic Logic*, vol. 21, pp. 253-254, 1956.

[10] Blake, A., "Canonical expressions in Boolean algebra," Dissertation, Dept. of Mathematics, Univ. of Chicago, 1937. Published by Univ. of Chicago Libraries, 1938.

[11] Bochmann, D., "Boolean differential calculus. A survey," (in Russian), *Izv. Akad. Nauk SSSR Tech. Kibernet.*, no. 5, pp. 125-133, 1977. English translation: *Engrg. Cybernet.*, vol. 15, no. 5, pp. 68-75.

[12] Boole, George, *The Mathematical Analysis of Logic*. London: G. Bell, 1847 (Reprinted by Philosophical Library, New York, 1948).

[13] Boole, George, *An Investigation of the Laws of Thought*. London, Walton, 1854 (Reprinted by Dover Books, New York, 1954).

[14] Borland International, *Turbo Pascal Owner's Handbook*, Scotts Valley, CA, 1987.

[15] Bossen, D.C. & S.J. Hong, "Cause-effect analysis for multiple fault detection in combinational networks," *IEEE Trans. on Computers*, vol. C-20, pp. 1252-1257, Nov. 1971.

[16] Brand, D., "Logic Synthesis," in *Design Systems for VLSI Circuits*, ed. by G. De Micheli, A. Sangiovanni-Vincentelli, and P. Antognetti. Boston: Martinus Nijhoff Publishers, 1987.

[17] Brayton, R.K. & C. McMullen, "The decomposition and factorization of Boolean expressions," Proc. Int'l. Symp. on Circuits and Systems, pp. 49-54, 1982.

[18] Brayton, R.K., G.D. Hachtel, C.T. McMullen, and A.L. Sangiovanni-Vincentelli, *Logic Minimization Algorithms for VLSI Synthesis*. Boston: Kluwer Academic Publishers, 1984.

[19] Brayton, R.K., "Factoring logic functions," IBM J. Res. Develop., vol. 31, no. 2, pp. 1877-198, March 1987.

[20] Brayton, R.K.,"Algorithms for Multi-Level Logic Synthesis and Optimization," in *Design Systems for VLSI Circuits*, ed. by G. De Micheli, A. Sangiovanni-Vincentelli, and P. Antognetti. Boston: Martinus Nijhoff Publishers, 1987.

[21] Bredeson, J.G. and P.T. Hulina, "Generation of prime implicants by direct multiplication," *IEEE Trans. on Computers*, vol. C-20, pp. 475-476, 1971.

[22] Breuer, M.A., S.J. Chang, and S.Y.H. Su, "Identification of multiple stuck-type faults in combinational networks," *IEEE Transactions on Computers*, vol. C-25, no. 1, pp. 44-54, January 1976.

[23] Brown, F.M., "Reduced solutions of Boolean equations," *IEEE Trans. on Computers*, vol. C-19, pp. 976-981, 1970.

[24] Brown, F.M., "Single-parameter solutions of flip-flop equations," *IEEE Trans. on Computers*, vol. C-20, pp. 452-454, April, 1971.

[25] Brown, F.M., "On a convenient division of labor in the generation of prime implicants," *Computers and Electrical Engineering*, vol. 6, pp. 267-271, 1979.

[26] Brown, F.M. and S. Rudeanu, "Consequences, consistency and independence in Boolean algebras," *Notre Dame J. Formal Logic*, vol. 22, no. 1, pp. 45-62, 1981.

[27] Brown, F.M., "Segmental solutions of Boolean equations," *Discrete Applied Mathematics*, vol. 4, pp. 87-96, 1982.

[28] Brown, F.M. and S. Rudeanu, "Recurrent covers and Boolean equations," Proc. Colloq. on Lattice Theory, Szeged, Hungary, Aug. 1980. Published in *Colloquia Mathematica Societatis Janos Bolyai*, North-Holland Pub. Co., vol. 33, pp. 55-86, 1983.

[29] Brown, F.M. and S. Rudeanu, "Prime implicants of dependency functions," *Analele Universității București*, vol. 37, no. 2, pp. 16-11, 1988.

[30] Brzozowski, J.A. and M. Yoeli, *Digital Networks*. Englewood Cliffs, NJ: Prentice-Hall, 1976.

[31] Bunitskiy, E., "Some applications of mathematical logic to the theory of the greatest common divisor and least common multiple" (in Russian), *Vestnik Opytnoy fiziki i elem. mat.*, no. 274, 1899.

[32] Burgoon, R., "Improve your Karnaugh mapping skills," *Electronic Design*, 21 December 1972, pp. 54-56.

[33] Caldwell, S.H., *Switching Circuits and Logical Design*. New York: Wiley, 1958.

[34] Carroll, L., *Symbolic Logic*. (Fourth Edition) London, 1896 (reprinted by Dover Publications, 1958).

[35] Carvallo, M., *Principes et Applications de l'Analyse Booléenne*. Paris: Gauthier-Villars, 1965.

[36] Cerny, E. and M.A. Marin, "An approach to unified methodology of combinational switching circuits," *IEEE Trans. Comput.*, vol. C-26, no. 8, pp. 745-756, August 1977.

[37] Chang, D.M.Y. and T.H. Mott, "Computing irredundant normal forms from abbreviated presence functions," *IEEE Trans. on Computers*, vol. EC-14, pp. 335-342, June, 1965.

[38] Chang, C.L. and R.C.T. Lee, *Symbolic Logic and Mechanical Theorem Proving*. New York: Academic Press, 1973.

[39] Clare, C.R., *Designing Logic Systems Using State Machines*. New York: McGraw-Hill, 1973.

[40] Clocksin, W.F. and C.S. Mellish, *Programming in Prolog*. New York: Springer-Verlag, 1981.

[41] Couturat, L., *L'algèbre de la Logique*. Paris: Scientia, 1905. English translation (by Lydia G. Robinson): Open Court Pub. Co., Chicago & London, 1914.

[42] Curtis, H.A., *A New Approach to the Design of Switching Circuits*. Princeton, N.J.: Van Nostrand, 1962.

[43] Cutler, R.B. and S. Muroga, "Derivation of minimal sums for completely specified functions," *IEEE Trans. Comput.*, vol. C-36, no. 3, pp. 277-292, March 1987.

[44] Darringer, J.A., Joyner, W., Berman, L. & Trevillyan, L., "Logic synthesis through local transformations," IBM J. of R. and D., vol. 25, pp. 272-280, July 1981.

[45] Davio, M. and J.-P. Deschamps, "Classes of solutions of Boolean equations, *Philips Research Reports*, vol. 24, pp. 373-378, October 1969.

[46] Davio, M., J.-P. Deschamps and A. Thayse, *Discrete and Switching Functions*. New York: McGraw-Hill, 1978.

[47] Davis, M. and H. Putnam, "A computing procedure for quantification theory," *J. Assoc. for Computing Machinery*, vol. 7, pp. 201-215, 1960.

[48] Delobel, C. and R.G. Casey, "Decomposition of a data base and the theory of Boolean switching functions," *IBM J. Res. & Develop.*, vol. 17, pp. 374-386, 1973.

[49] Deschamps, J.P., "Maximal classes of solutions of Boolean equations," *Philips Research Reports*, vol. 26, pp. 249-260, August 1971.

[50] Dietmeyer, D.L., *Logic Design of Digital Systems*, Second Edition. Boston: Allyn & Bacon, 1978.

[51] Dunham, B., R. Fridshal, and G.L. Sward, "A nonheuristic program for proving elementary logical theorems," *Proc. Int'l. Conf. on Inf. Processing* (Paris: UNESCO), 1959, pp. 282-284.

[52] Dunham, B. and J.H. North, "Theorem testing by computer," *Symposium on Mathematical Theory of Automata*, Polytechnic Inst. of Brooklyn, 1962.

[53] Dunham, B. and H. Wang, "Towards feasible solutions to the tautology problem," *Ann. Math. Logic*, vol. 10, pp. 117-154, 1976.

[54] Ehrenfest, P., "Review of L. Couturat, 'The Algebra of Logic'," *Journ. Russian Phys. & Chem. Soc., sec. 2*, vol. 42, no. 10, p. 382, 1910.

[55] Elgot, C.C., *Lectures on Switching and Automata Theory*, Technical Report, University of Michigan, Ann Arbor, Mich., Jan. 1959.

[56] Ewing, A.C. et al., "Algorithms for logical design," *Comm. & Electronics*, no. 56, pp. 450-458, 1961.

[57] Fletcher, W.I., *An Engineering Approach to Digital Design*, Englewood Cliffs, NJ: Prentice-Hall, 1980.

[58] Florine, J., "Optimization of binary functions with a special-purpose electronic computer," *Automation and Remote Control*, vol. 28, pp. 956-962, 1967.

[59] Florine, J., *The Design of Logical Machines*. New York: Crane, Russak & Co., 1973.

[60] Frege, G., *Begriffsschrift, Eine Der Arithmetischen Formalsprache Des Reinen Denkens*. Halle: Nebert, 1879 (Translated in [207]).

[61] Friedman, A.D., *Logical Design of Digital Systems*. Woodland Hills, CA: Computer Science Press, 1975.

[62] Galil, Z., "The complexity of resolution procedures for theorem proving in the propositional calculus," Department of Computer Science, Cornell University, TR 75-239, 1975.

[63] Gann, D., J.D. Schoeffler, and L.E. Ostrander, "A finite-state model for the control of adrenal cortical steroid secretion," in M.D. Mesarović (Ed.), *Systems Theory and Biology*. New York: Springer-Verlag, 1968.

[64] Gardner, M., *Logic Machines and Diagrams*. McGraw-Hill, 1958.

[65] Garey, M.R. and D.S. Johnson, *Computers and Intractability*. San Francisco: W.H. Freeman, 1979.

[66] Gavrilov, M.A. and A.D. Zakrevskii (Ed's.), *LYaPAS: A Programming Language for Logic and Coding Algorithms*. NY: Academic Press, 1969.

[67] Genesereth, M.R., "The role of design descriptions in automated diagnosis," *Artificial Intelligence*, vol. 24, pp. 411-436, Dec. 1984.

[68] Genesereth, M.R. and M.L. Ginsberg, "Logic Programming," *Communications of the ACM*, vol. 28, no. 9, Sept. 1985.

[69] Genesereth, M.R. and N.J. Nilsson, *Logical Foundations of Artificial Intelligence*. Los Altos, CA: Morgan Kaufmann, 1987.

[70] Ghazala, M.J. "Irredundant disjunctive and conjunctive forms of a Boolean function," *I.B.M. Journal of Research and Development*, vol. 1, pp. 171-176, April 1957.

[71] Givone, D.G., *Introduction to Switching Circuit Theory*. New York: McGraw-Hill, 1970.

[72] Goodstein, R.L., *Boolean Algebra*. New York: Macmillan, 1963.

[73] Gómez-González, L., *Estudio teorico, concepcion y realizacion de un sistema electronico para simplificar funciones logicas*, Dissertation, Dpto. Electricidad y Electronica, Facultad de Ciencias, Universidad de Granada, Spain, 1977.

[74] Gray, F., "Pulse Code Communication," U.S. Patent 2,632,058, 17 Mar., 1953.

[75] Grinshpon, M.S.,"Selection criterion for a potentially inessential argument to be eliminated from an incompletely-specified logical function," *Automatic Control and Computer Sciences* vol. 9, no. 5, pp. 16-18 (translated from *Automatika i Vychislitel'naya Tekhnika*, USSR), 1975.

[76] Halatsis, C. and N. Gaitanis, "Irredundant normal forms and minimal dependence sets of a Boolean function," *IEEE Trans. on Computers,*, vol. C-27, no. 11, pp. 1064-1068, Nov. 1978.

[77] Halmos, P.R., *Naive Set Theory.* Princeton, N.J.: D. Van Nostrand Co., 1960.

[78] Halmos, P.R., *Lectures on Boolean Algebras.* New York: Springer-Verlag, 1974.

[79] Hammer, P.L. and S. Rudeanu, *Boolean Methods in Operations Research.* New York: Springer-Verlag, 1968.

[80] Harrison, M.A., *Introduction to Switching and Automata Theory.* New York: McGraw-Hill, 1965.

[81] Hartmanis, J., "Symbolic analysis of a decomposition of information processing machines," *Information and Control*, vol. 3, no. 2, pp. 154-178, June 1960.

[82] Harvard Computation Laboratory Staff, *Synthesis of Electronic Computing and Control Circuits*, Annals of the Computation Lab., vol. 27. Cambridge, Mass.: Harvard Univ. Press, 1951. Chapter VII, "Multiple-output circuits."

[83] Hight, S.L.,"Minimal input solutions," *IEEE Trans. on Computers,*, vol. C-20, no. 8, pp. 923-925, Aug. 1971.

[84] Hill, F.J. and G.R. Peterson, *Switching Theory and Logical Design*, Third Edition. New York: Wiley, 1981.

[85] Ho, B., "NAND synthesis of multiple-output combinational logic using implicants containing output variables," Ph.D. Dissertation, U. of Wisconsin, 1976.

[86] Hohn, F., *Applied Boolean Algebra*. Second Edition. New York & London: Macmillan, 1966.

[87] Horowitz, I.A., *Chess for Beginners*. Irvington-on-Hudson, N.Y.: Capitol Publ. Co., 1950.

[88] House, R.W. and T. Rado, "A generalization of Nelson's algorithm for obtaining prime implicants," *J. Symb. Logic*, vol. 30, pp. 8-12, 1965.

[89] Huffman, D.A., "Solvability criterion for simultaneous logical equations," M.I.T. Research Lab. of Electronics, Quarterly Progress Report No. 48, AD 156-161, 15 Jan. 1958.

[90] Huffman, D.A., "Combinational circuits with feedback," Chapter 2 of *Recent Developments in Switching Theory* (ed. A. Mukhopadhyay), pp. 27-55, Academic Press, N.Y., 1971.

[91] Hulme, B.L. and R.B. Worrell, "A prime implicant algorithm with factoring," *IEEE Trans. on Computers*, vol. C-24, pp. 1129-1131, 1975.

[92] Huntington, E.V., "Sets of independent postulates for the algebra of logic," *Trans. Amer. Math. Soc.*, vol. 5, pp. 288-309, 1904.

[93] Jesse, J.E., "A more efficient use of Karnaugh Maps," *Computer Design*, February 1972, pp. 80-82.

[94] Jevons, W.S., *Pure Logic, or the Logic of Quality Apart from Quantity*. London: Stanford, 1864.

[95] Kabat, W.C. and A.S. Wojcik, "Automated synthesis of combinational logic using theorem-proving techniques," *Proc. Twelfth Int'l. Symp. on Multiple-Valued Logic*, pp. 178-199, (May 1982); *IEEE Trans. Computers*, vol. C-34, no. 7, pp. 610-632, July 1985.

[96] Kainec, James J., "A diagnostic system using Boolean reasoning," M.S. Thesis, Air Force Institute of Technology, Wright-Patterson AFB, Ohio, December 1988.

[97] Kalish, D. and R. Montague, *Logic: Techniques of Formal Reasoning*. New York: Harcourt Brace Jovanovich, 1964.

[98] Kambayashi, Y., "Logic design of programmable logic arrays," *IEEE Trans. on Computers*, vol. C-28, pp. 609-617, Sept. 1979.

[99] Karnaugh, M., "The map method for synthesis of combinational logic circuits," *AIEE Trans. on Comm. & Electronics*, vol. 9, pp. 593-599, 1953.

[100] Kautz, W.H., "The necessity of closed circuit loops in minimal combinational circuits," IEEE Trans. on Computers, vol. C-19, no. 2, pp. 162-164, Feb. 1970.

[101] Keynes, J.N., *Studies and Exercises in Formal Logic*, Second Edition. London: Macmillan, 1887.

[102] Kjellberg, G. "Logical and other kinds of independence," *Proc. of an Int'l. Symp. on the Theory of Switching, Annals of the Computer Lab. of Harvard U.*, vol. 39, Part I, pp. 117-124, Harvard U. Press, 1959.

[103] Klir, G.J. and M.A. Marin, "New considerations in teaching switching theory," *IEEE Trans. on Education*, vol. E-12, pp. 257-261, 1969.

[104] Klir, G.J., *Introduction to the Methodology of Switching Circuits*. New York: D. Van Nostrand Co., 1972.

[105] Kobrinsky, N.E. & Trakhtenbrot, B.A., *Introduction to the Theory of Finite Automata*. Amsterdam: North-Holland Publ. Co., 1965. Chapter VI, Section 3, "Synthesis of a multi-output logical net."

[106] Kohavi, Z., *Switching and Finite Automata Theory*. New York: McGraw-Hill, 1970.

[107] Korfhage, R.R., *Logic and Algorithms, With Applications to the Computer and Information Sciences*. New York: Wiley, 1966.

[108] Kowalski, R., *Logic for Problem Solving*. Amsterdam, New York: North-Holland, 1979.

[109] Krieger, M., *Basic Switching Circuit Theory*. New York: Macmillan, 1967.

[110] Kuntzmann, J., *Algèbre de Boole*. Paris: Dunod, 1965.

[111] Ladd, Christine, "On the algebra of logic," in *Studies in Logic*, ed. by C. S. Peirce. Boston: Little, Brown & Co., 1883, pp. 17-71.

[112] Lazarev, V.G. and E.I. Piil', "On the integration of potential-pulse forms," Soviet Physics – Doklady, vol. 6, no. 7, 1962.

[113] Ledley, R.S., "A digitalization, systematization, and formulation of the theory and methods of the propositional calculus," NBS Report 3363,

Nat'l. Bureau of Standards, U.S. Dep't. of Commerce, (U.S. Gov't, document no. AD56-412), 1 Feb. 1954.

[114] Ledley, R.S., "Mathematical foundations and computational methods for a digital logic machine," *J. Ops. Res. Soc. Amer.*, vol. 2, pp. 249-274, 1954.

[115] Ledley, R.S., "Digital computational methods in symbolic logic, with examples in biochemistry," *Proc. Nat'l. Acad. Sci.*, vol. 41, pp. 498-511, July 1955.

[116] Ledley, R.S., "Logical aid to systematic medical diagnosis (and operational simulation in medicine)," *J. Ops. Res. Soc. Amer.*, vol. 4, no. 3, p. 392, Aug. 1956.

[117] Ledley, R.S. and L.B. Lusted, "Reasoning foundations of medical diagnosis," *Science*, vol. 130, no. 3366, pp. 9-21, 3 July, 1959.

[118] Ledley, R.S., *Digital Computer and Control Engineering*. New York: McGraw-Hill Book Co, 1960.

[119] Ledley, R.S., *Use of Computers in Biology and Medicine*. New York: McGraw-Hill Book Co, 1965. Chapter 12, "Medical diagnosis and medical record-keeping."

[120] Lee, R.C.T., "An algorithm to generate prime implicants and its application to the selection problem," *Inf. Sciences*, vol. 4, pp. 251-254, July 1972.

[121] Lee, S.C., *Digital Circuits and Logic Design*. Englewood Cliffs, NJ: Prentice-Hall, 1976.

[122] Lee, S.C., *Modern Switching Theory and Digital Design*. Englewood Cliffs, NJ: Prentice-Hall, 1978.

[123] Lewis, C.I., *A Survey of Symbolic Logic*. Berkeley: U. of Cal. Press, 1918. Reprinted by Dover Pub's., Inc., New York, 1960. Chapt. II, "The Classic, or Boole-Schröder Algebra of Logic."

[124] Löwenheim, L., "Uber die Auflösung von Gleichungen im logischen Gebietekalkul," *Math. Ann*, vol. 68, 1910, pp. 169-207. Translation: "The solution of equations in the calculus of logic," AFCRL-69-0149, Air Force Cambridge Research Laboratories, April, 1969.

[125] Luckham, D., "The resolution principle in theorem-proving," in *Machine Intelligence 1* (N.L. Collins and D. Michie, ed's.), Edinburgh & London: Oliver & Boyd, 1967.

[126] Maghout, K., "Détermination des nombres de stabilité et du nombre chromatique d'un graphe." *C. R. Acad. Sci. Paris*, vol. 248, pp. 3522-23, 1959.

[127] Maghout, K., "Applications de l'algèbra de Boole à la théorie des graphes et aux programmes linéaires et quadratiques," *Cahiers Centre Edudes Réch. Opér.*, vol. 5, pp. 21-99, 1963.

[128] Marczewski, E., "Independence in algebras of sets and Boolean algebras," *Fundamenta Mathematicae*, vol. 48, pp. 135-145, 1960.

[129] Marcus, M.P., "Derivation of maximal compatibles using Boolean algebra," *I.B.M. J. Res. & Devel.*, vol. 8, pp. 537-538, 1964.

[130] Marin, M.A., "Investigation of the field of problems for the Boolean Analyzer," Report No. 68-28, Dep't. of Engineering, U. of Calif. at Los Angeles, 1968.

[131] Marquand, A., "A logical diagram for n terms," *Philosophical Magazine*, vol. 12, pp. 266-270, 1881.

[132] May, A., "Adaptive location of multiple faults in combinational circuits," M.S. thesis, Department of Electrical Engineering, University of Kentucky, Lexington, KY, August, 1984.

[133] McCaw, C.R., "Loops in directed combinational switching circuits," Stanford Electronics Lab's., T.R. No. 6208-1, April 1963.

[134] McCluskey, E.J., "Minimization of Boolean functions," *Bell Sys. Tech. J.*, vol. 35, pp. 1417-1444, 1956.

[135] McColl, H., "The calculus of equivalent statements," *Proc. London Math. Soc.*, vol. 9 (1877/78), pp. 9-20; vol. 10 (1878), pp. 16-28; vol. 11 (1879/80), pp. 113-121.

[136] McCluskey, E.J., *Introduction to the Theory of Switching Circuits.* New York: McGraw-Hill, 1965.

[137] Mendelson, E., *Boolean Algebra and Switching Circuits.* New York: McGraw-Hill (Schaum's Outline Series), 1970.

[138] Mitchell, O.H., "On a new algebra of logic," in *Studies in Logic*, ed. by C.S. Peirce. Boston: Little, Brown, & Co, 1883.

[139] Mithani, D., "Implementation of NAND synthesis using implicants containing output variables," M.S. thesis, Dep't. of Electrical Engineering, Univ. of Wisconsin, 1977.

[140] Mott, T.H., "Determination of the irredundant normal forms of a truth function by iterated consensus of the prime implicants," *IRE Trans. on Electronic Computers*, vol. EC-9, pp. 245-252, June 1960.

[141] Muller, D.E., "Application of Boolean algebra to switching circuit design and to error detection," *Trans. IRE*, vol. EC-3, pp. 6-12, Sept. 1954.

[142] Müller, E., *Abriss der Algebra der Logik, 1909-10*. (Appendix to vol. III of [178]).

[143] Muroga, S., *Logic Design and Switching Theory*. New York: Wiley-Interscience, 1979.

[144] Naito, S., "Algebraic analysis for asynchronous sequential circuits," *NEC Research and Development*, No. 34, pp. 80-89, July 1974.

[145] Nakasima, A., "The theory of equivalent transformation of simple partial paths in relay circuits" (in Japanese), *J. Inst. Elec. Commun. Engrs. Japan*, no. 165, 167, Dec. 1936, Feb. 1937.

[146] Nakasima, A., "Algebraic expressions relative to simple partial paths in the relay circuit" (in Japanese), *J. Inst. Electrical Communication Engineers of Japan*, no. 173, August 1937 (condensed English translation: *Nippon Electrical Comm. Engineering*, no. 12, pp. 310-314, Sept. 1938). Section V, "Solutions of acting impedance equations of simple partial paths."

[147] Nelson, R.J., "Simplest normal truth functions," *J. Symb. Logic*, vol. 20, pp. 105-108, 1955.

[148] Nelson, R.J., *Introduction to Automata*. New York: Wiley, 1968.

[149] Nilsson, N.J., *Problem-Solving Methods in Artificial Intelligence*. New York: McGraw-Hill, 1971. Chapter 6: "Theorem-Proving in the Predicate Calculus."

[150] Nilsson, N.J., *Principles of Artificial Intelligence*. Palo Alto, Calif.: Tioga Publ. Co., 1980.

[151] Peirce, C.S., "On the algebra of logic," *Amer. J. of Math.*, vol. 3, pp. 15-57, 1880.

[152] Peirce, C.S., ed., *Studies in Logic*. By Members of the Johns Hopkins University. Boston: Little Brown & Co, 1883.

[153] Peirce, C.S., "Logical machines," *Amer. J. Psychology*, vol. 1, pp. 165-170, 1887.

[154] Petrick, S.R., "A direct determination of the irredundant forms of a Boolean function from a set of prime implicants," A.F. Cambridge Res. Center, Bedford, Mass., Report AFCRC-TR-56-110, 1956.

[155] Phister, M., *Logical Design of Digital Computers*. New York: John Wiley, 1958.

[156] Pichat, E., "Algorithms for finding the maximal elements of a finite universal algebra," *Information Processing 68, Proc. IFIP Congress*, pp. 214-218, 1968.

[157] Poage, J.F., "Derivation of optimum tests to detect faults in combinational circuitry," *Mathematical Theory of Automata*, MRI Symposium Series, Volume XII, Polytechnic Institute of Brooklyn, 1963

[158] Poretsky, P., "On methods for solving logical equations and on the inverse method for mathematical logic" (in Russian), *Bull. de la Soc. Physico-Mathématique de Kasan*, vol. 2, pp. 161-130, 1884.

[159] Poretsky, P., "Sept lois fondamentales de la théorie des égalités logiques," *Bull. de la Soc. Physico-Mathématique de Kasan*, ser. 2, vol. 8, pp. 33-103, 129-181, 183-216, 1898.

[160] Pratt, W.C., "Transformation of Boolean equations for the design of multiple-output networks," Dissertation, Electrical Engrg. Department, University of Illinois, 1976.

[161] Quine, W.V., "The problem of simplifying truth functions," *Am. Math. Monthly*, vol. 59, pp. 521-531, 1952.

[162] Quine, W.V., "Two theorems about truth functions," *Bol. Soc. Math. Mexicana*, vol. 10, pp. 64-70, 1953.

[163] Quine, W.V., "A way to simplify truth functions," *Am. Math. Monthly*, vol. 62, pp. 627-631, 1955.

[164] Quine, W.V., "On cores and prime implicants of truth functions," *Am. Math. Monthly*, vol. 66, pp. 755-760, 1959.

[165] Reed, I.S., "A class of multiple error-correcting codes and the decoding scheme," *IRE Trans. on Information Theory*, vol. IT-4, pp. 38-49, Sept. 1954.

[166] Reusch, B., "Generation of prime implicants from subfunctions and a unifying approach to the covering problem," *IEEE Trans. on Computers*, vol. C-24, no. 9, pp. 924-930, September 1975.

[167] Reusch, B. and L. Detering, "On the generation of prime implicants," *Annales Societatis Mathematicae Polonae, Series IV: Fundamenta Informaticae II*, pp. 167-186, 1979.

[168] Robinson, J.A., "A machine oriented logic based on the resolution principle," *Journal of the Association for Computing Machinery*, vol. 12, no. 1, pp. 23-41, January 1965.

[169] Rose, A., *Computer Logic*. New York: Wiley-Interscience, 1971.

[170] Rosenbloom, P., *The Elements of Mathematical Logic*. New York: Dover Publications, 1950.

[171] Rudeanu, S., "Boolean equations and their applications to the study of bridge circuits. I," *Bull. Math. Soc. Math. Phys. R. P. Roumaine*, vol. 3, pp. 445-473, 1959.

[172] Rudeanu, S., *Boolean Functions and Equations*. Amsterdam-London-New York: North-Holland Publ. Co. & American Elsevier, 1974.

[173] Rushdi, A.M., "Improved variable-entered Karnaugh map procedures," Computers and Electrical Engineering, vol. 13, no. 1, pp. 41-52, 1987.

[174] Samson, E.W. and B.E. Mills, "Circuit minimization: algebra and algorithms for new Boolean canonical expressions," Air Force Cambridge Research Center, AFCRC TR 54-21, April, 1954.

[175] Samson, E.W. and R.K. Mueller, "Circuit minimization: sum to one process for irredundant sums," Air Force Cambridge Research Center, Report AFCRC-TR-55-118, August 1955.

[176] Sasao, T., "HART: a hardware for logic minimization and verification," *Internat'l. Conf. on Computer-Aided Design, ICCD-85*, pp. 713-718, 1985.

[177] Schoeffler, J.D., L.E. Ostrander, and D.S. Gann, "Identification of Boolean mathematical models," in M.D. Mesarović (Ed.), *Systems Theory and Biology*. New York: Springer-Verlag, 1968.

[178] Schröder, E., *Vorlesungen über die Algebra der Logik*. Leipzig: Vol. 1, 1890; Vol. 2, 1891; Vol. 3, 1895; Vol. 2, Part 2, 1905. Reprint: Chelsea Pub. Co., Bronx, N.Y., 1966.

[179] Schultz, G.W., "An algorithm for the synthesis of complex sequential networks," *Computer Design*, March, 1969, pp. 49-55.

[180] Sellers, F.F., M.Y. Hsiao and L.W. Bearnson, "Analyzing errors with the Boolean difference," *IEEE Trans. Computers*, vol. C-17.7, pp. 676-683, July 1968.

[181] Semon, W., "The application of matrix methods in the theory of switching," Doctoral thesis, Comp. Lab., Harvard Univ., Cambridge, Mass., April 1954.

[182] Semon, W., "A class of Boolean equations," Report SRRC-RR-17, Sperry Rand Research Center, Sudbury, Mass., 1962.

[183] Shannon, C.E., "A symbolic analysis of relay and switching circuits," *Trans. Amer. Inst. Elec. Engrs.*, vol. 57, pp. 713-723, 1938.

[184] Shannon, C.E., "The synthesis of two-terminal switching circuits," *Bell System Tech. J.*, vol. 28, no. 1, pp. 59-98, 1949.

[185] Shestakov, V.I., "Some mathematical methods for construction and simplification of two-terminal electrical networks of class A" (in Russian), Dissertation, Lomonosov State University, Moscow, 1938.

[186] Short, R.A., "A theory of relations between sequential and combinational realizations of switching functions," Stanford Electronics Laboratories, T.R. No. 098-1, 12 Dec., 1960.

[187] Sikorski, R., *Boolean Algebras*. New York: Springer-Verlag, 1969.

[188] Slagle, J.R., et al., "A new algorithm for generating prime implicants," *IEEE Trans. on Computers*, vol. C-19, pp. 304-310, 1970.

[189] Small, A.W., "A new approach to functional decomposition," Air Force Cambridge Research Laboratories, Report AFCRL-71-0010, 28 Dec., 1970.

[190] Stone, M.H., "The theory of representations for Boolean algebras," *Trans. Amer. Math. Soc.*, vol. 40, pp. 37-111, 1936.

[191] Svoboda, A., "Boolean analyzer," *Information Processing 68 (Proc. IFIP Congress, Edinburgh)*. Amsterdam: North-Holland, pp. 824-830, 1969.

[192] Svoboda, A., "Parallel processing in Boolean algebra," *IEEE Trans. on Computers*, vol. C-22, pp. 848-851, 1973.

[193] Svoboda, A. and D.E. White, *Advanced Logical Circuit Design Techniques*. New York: Garland STPM Press, 1979.

[194] Talantsev, A.D., "On the analysis and synthesis of certain electrical circuits by means of special logical operators," *Automation and Remote Control*, vol. 20, no. 9, pp. 874-883, 1959.

[195] Tapia, M.A., J.H. Tucker and A.W. Bennett, "Boolean integration," *Proc. IEEE Southeast-Con*, Clemson, SC, April 1976.

[196] Tapia, M.A., J.H. Tucker and A.W. Bennett, "Boolean differentiation and integration using Karnaugh Map," *Proc. IEEE Southeast-Con*, 1977.

[197] Tapia, M.A., "Application of Boolean calculus to digital system design," *Proc. IEEE Southeast-Con*, Nashville, Tenn., 14-16 April, 1980.

[198] Tapia, M.A. and J.H. Tucker, "Complete solution of Boolean equations," *IEEE Trans. on Comput.*, vol. C-29, no. 7, pp. 662-665, July 1980.

[199] Tapia, M.A. "Boolean integral calculus for digital systems," *IEEE Trans. on Comput.*, vol. C-34, no. 1, pp. 78-81, Jan. 1985.

[200] Taylor, D.K., "Analyzing Relational Databases using Propositional Logic," M.S. Thesis, Department of Electrical Engineering, University of Kentucky, December, 1981.

[201] Texas Instruments, Inc., *The TTL Data Book for Design Engineers*, 1973.

[202] Thayse, A., "Boolean differential calculus," *Philips Res. Rept's.*, vol. 26, pp. 229-246, 1971.

[203] Thayse, A. and M. Davio, "Boolean differential calculus and its applications in switching theory," *IEEE Trans. Comput.*, vol. C-22, pp. 409-420, 1973.

[204] Tison, P., *Theorie des consensus*, Dissertation, University of Grenoble, France, 1965.

[205] Tison, P., "Generalization of consensus theory and application to the minimization of Boolean functions," *IEEE Trans. Electronic Computers*, vol. EC-16, pp. 446-456, 1967.

[206] Uehara, T. and N. Kawato, "Logic circuit synthesis using Prolog," *New Generation Computing*, vol. 1, no. 2, 1983.

[207] van Heijenoort, J. (Ed.), *From Frege To Gödel: A Source Book Of Mathematical Logic, 1897-1931*. Cambridge, Mass.: Harvard University Press, 1967.

[208] Veitch, E.W., "A chart method for simplifying truth functions," *Proc. ACM Conference*, Pittsburgh, Pa., 2-3 May, 1952, pp. 127-133.

[209] Venn, J., "On the employment of geometrical diagrams for the sensible representation of logical propositions," *Proc. Cambridge Philosophical Society*, vol. 4, pp. 35-46, 1880.

[210] Venn, J., *Symbolic Logic, 2nd edition*. London, Macmillan, 1894. (Reprinted by Chelsea Pub. Co., New York, 1971).

[211] Weissman, J., "Boolean algebra, map coloring and interconnections," *Amer. Math. Monthly*, vol. 69, pp. 606-613, 1962.

[212] Whitehead, A.N., *A Treatise on Universal Algebra, with Applications*. Cambridge: The University Press, 1898.

[213] Whitehead, A.N., "Memoir on the algebra of symbolic logic, Part I," *Am. J. of Math.*, vol. 23, pp. 139-165, 297-316, 1901.

[214] Whitesitt, J.E., *Boolean Algebra and its Applications*. Reading, MA: Addison-Wesley, 1961.

[215] Wojciechowski, W.S. and A.S. Wojcik, "Multiple-valued logic design by theorem proving," *Proc. Ninth. Int'l. Symp. on Multiple-Valued Logic*, Bath, England, 1979, pp. 196-199.

[216] Wojciechowski, W.S., *Multiple-valued combinational logic design using theorem proving*. Dissertation, Ill. Inst. of Tech., 207 pp. University Microfilms No. KRA80-2162, May 1980.

[217] Wojciechowski, W.S. and A.S. Wojcik, "Automated design of multiple-valued logic circuits by automated theorem-proving techniques," *IEEE Trans. on Computers*, vol. C-32, pp. 785-798, Sept. 1983.

[218] Wood, P.E., Jr., *Switching Theory*. New York: McGraw-Hill, 1968.

[219] Wos, L., R. Overbeek, E. Lusk & J. Boyle, *Automated Reasoning: Introduction And Applications*. Englewood Cliffs, N.J.: Prentice-Hall, 1984.

[220] Yamada, K. and K. Yoshida, "An application of Boolean algebra in practical situations," *Hitotsubashi J. Arts & Sciences*, vol. 5, pp. 41-57, 1965.

[221] Zakrevskii, A.D. and A.Yu. Kalmykova, "The solution of systems of logical equations," in [66], pp. 193-206.

[222] Zakrevskii, A.D., "Testing for identities in Boolean algebra," in [66], pp. 207-213.

[223] Zhegalkin, I.I., "On the calculation of propositions in symbolic logic," (in Russian), *Math. Sbornik*, vol. 34, pp. 9-28, 1927.

Index

1. **The specification must be tabular.** The restriction to tabular specifications is not as serious as it might appear. Essentially all existing design-techniques assume a tabular specification; thus no novelty in specification is introduced. A non-tabular specification can be handled by decomposing it into a collection of tabular specifications; a solution of any of the component tabular specifications is a solution of the original non-tabular specification. Methods for carrying out such decomposition are discussed in [27].

2. **Cost is measured by gate-inputs.** The cost of a recursive solution is defined (in the program whose operation is described in the next subsection) to be *gate-input count* (*cf.* Section 9.5). This cost-function measures the number of inputs to gates in a two-level (AND-to-OR or NAND-to-NAND) realization of the formula, assuming that the complemented input-signals $x_1', x_2', \ldots, x_m'$ are available. The cost of a solution is the sum of the costs of its component formulas.

3. **Feedback is excluded.** The organization of a recursive solution excludes closed loops, thereby guaranteeing that the corresponding circuit is strongly combinational. Kautz [100] has shown that such loops may be necessary to achieve minimal cost in a combinational circuit-design, and Pratt [160] has developed transformation-techniques that produce strongly combinational designs incorporating closed loops. Restricting ourselves to recursive solutions, however, greatly reduces computational complexity. This reduction is gained, we believe, without significant increase in gate-input cost.

4. **Redundant variables are excluded.** The candidate argument-sets to generate output $z_i \in Z$, for any value of i, are the *minimal determining subsets* of $\mathcal{T}(z_i)$. There are cases in which additional (and logically superfluous) arguments are needed to attain minimal cost; such a case is exhibited in Example 9.6.3. Such cases seem rare and the cost-advantage to be gained by introducing superfluous arguments seems minor; the exclusive use of minimal determining subsets is therefore justified by the drastic reduction it induces in the space of formulas to be searched.